HYDRA 2000

Proceedings of the XXVIth Congress of the International Association for Hydraulic Research, London, 11–15 September 1995.

5. John F. Kennedy student paper competition

Thomas Telford

International conference organized by the International Association for Hydraulic Research and the Institution of Civil Engineers

Published by Thomas Telford Services Ltd, Thomas Telford House, 1 Heron Quay, London E14 4JD

First published 1995

Distributors for Thomas Telford books are
USA: American Society of Civil Engineers, Publications Sales Department, 345 East 47th Street, New York, NY 10017-2398
Japan: Maruzen Co. Ltd, Book Department, 3–10 Nihonbashi 2-chome, Chuo-ku, Tokyo 103
Australia: DA Books and Journals, 648 Whitehorse Road, Mitcham 3132, Victoria

A CIP catalogue record for this book is available from the British Library.

Classification
Availability: unrestricted
Content: collected papers
Status: refereed papers
User: civil and hydraulic engineers

ISBN: 0 7277 2060 0

Printed and bound in Great Britain by Redwood Books, Trowbridge, Wiltshire.

Preface

The International Association of Hydraulic Research (IAHR), founded in 1935, is a world wide organisation of engineers and scientists interested in hydraulics (including hydrology and fluid mechanics) and its practical application, including industrial and environmental aspects. IAHR stimulates and promotes both basic and applied research and strives to have science and technology contribute to the optimisation of world water management and industrial processes. Through its members, IAHR accomplishes its goal by a wide variety of activities, offering a common infrastructure for the hydraulic research community, one of the most important of these activities being the biennial congresses.

HYDraulics Research and its Application next century—HYDRA 2000—was the main theme of the twenty-sixth IAHR Congress held at the Institution of Civil Engineers in London from 11 to 15 September 1995. This congress theme was chosen so as to bring together researchers, hydraulics users and practitioners to expose and discuss matters of common interest.

The main theme was broken down into four topics:

1. **Integration of research approaches and applications.**
2. **Industrial hydraulics and multi-phase flows.**
3. **Future paths in maritime hydraulics.**
4. **The hydraulics of water resources and their development.**

The papers for each of these topics are produced in separate volumes of the proceedings. Additionally there are communications (short papers or technical notes), and student papers which were submitted for the John F. Kennedy Student competition.

At the Congress an invited lecture was given on each of the four topics by leading authorities. The extended papers on which these four lectures were based are also included in the proceedings, which thus provide an up-to-date account of the role of hydraulics research and the part it can play in the optimisation of future developments.

Not included in the proceedings are papers and discussion at five special seminars and two forums which formed an integral part of the Congress.

W. R. White
Chairman of the Local Organising Committee

Acknowledgements

In recent years the average attendance at IAHR congresses has exceeded 500, with worldwide participation. The organisation of an event of this size is a major task and involves many people. In the case of the London Congress I would particularly like to thank:

1. The Local Organising Committee:

> P. Ackers (Editorial Board), *Consultant*
> P. Avery, *British Hydromechanics Research Group*
> S. J. Darby (Finance), *National Rivers Authority*
> Dr D. A. Ervine, *Glasgow University*
> Prof. A. J. Grass, *University College London*
> Prof. P. Novak (Technical Programme), *Consultant*
> R. G. Purnell, *Ministry of Agriculture, Fisheries and Food*
> Prof. R. H. J. Sellin (Social Programme), *University of Bristol*
> R. L. Soulsby, *HR Wallingford*
> D. G. Wardle, *Consultant*
> Prof. B. B. Willetts, *University of Aberdeen*
> [Dr W. R. White (Chairman), *HR Wallingford*]

2. The Conference Office of the Institution of Civil Engineers:
   C. Chin, R. Coninx, B. O'Donoghue, S. Frye, N. Kerwood, J. Morris, S. Walton

3. The four editors of the Congress proceedings for their hard work in administering the review procedures and collating each volume.

4. The distinguished IAHR lecturer, Sir Geoffrey Palmer of the University of Iowa, USA.

5. The convenors of the seminars, the forums and the student competition.

6. All those 'behind-the-scenes' helpers including those who reviewed papers and the partners of most of the LOC members whose talents and energy were used extensively.

I would also like to thank the Institution of Civil Engineers for hosting the Congress and providing a degree of financial security to the conference organisers.

In addition I also gratefully acknowledge the support and sponsorship of the following:

> British Hydromechanics Research Group
> HR Wallingford Group Ltd
> Ministry of Agriculture, Fisheries and Food
> EMAP Business Communications
> Participants in the Exhibition
> Advertisers in the Exhibition Brochure

W. R. White
Chairman of the Local Organising Committee

Editor's preface

This volume of the proceedings of the XXVIth IAHR Congress consists of papers accepted as entries for the John F. Kennedy Student Paper Competition.

The student competition is a relatively recent feature of the biennial IAHR Congress and 1995 is the first time the competition has been named in honour of our past president, J. F. Kennedy. Entry to the competition was open to any undergraduate or postgraduate student with a university (or equivalent) registration at any time during 1993 to 1995 i.e. the period between the XXVth and XXVIth congresses. Students were invited to submit papers on any topic covered by IAHR—submissions being original not previously published work authored by students alone.

All the accepted papers were refereed by two independent reviewers. Considering the wide range of educational backgrounds represented by the students from eight countries and at different stages in their course of studies, the overall standard of the papers was judged as very good and can also be regarded as a contribution to the ongoing debate within IAHR on education and training of hydraulic engineers.

The first prize in the competition has been sponsored by the John F. Kennedy Student Paper Competition Award Fund and the second and third prizes by the BHR Group, UK. The quality of both the presentation at the Congress as well as the technical merit of the paper will be taken into consideration by an international panel in making the awards to be presented during the Congress.

As Convenor of the student competition I would like to thank all authors entering the competition, the reviewers and the sponsors of the competition prizes.

Emeritus Professor P. Novak
Convenor, John F. Kennedy Student Competition

Contents

NUMERICAL MODELLING OF TURBULENT FLOW IN COMPOUND CHANNELS BY THE FINITE ELEMENT METHOD

J. BORIS ABRIL C.

School of Civil Engineering, The University of Birmingham, England, UK

ABSTRACT

This paper describes a Finite Element-based numerical model to predict the lateral distribution of depth-averaged velocity and Reynolds shear stress as well as the boundary shear stress in compound channels. The model is validated against the analytical solution for the case of trapezoidal channels. A comparison of the model with SERC-FCF series 020501 experimental data is given as an example of a particular calibration philosophy.

INTRODUCTION

Flow structures in compound channels have been studied extensively over the past and present decade due to their importance in various hydraulic processes. Whereas one-dimensional models provide a general understanding of the flow processes, a more detailed description of these phenomena must be accomplished if further effects such as sediment transport are to be considered. The use of numerical-computational methods such as Finite Elements provides the possibility to develop more flexible and accurate models which help predict and understand these complex mechanisms. The purpose of this paper is to introduce a FE-based computer model which describes the lateral distribution of depth mean velocity in channels of any shape. The model is also able to compute lateral distributions of Reynolds and boundary shear stresses which are of foremost importance in analysing turbulence and sediment transport phenomena.

GOVERNING DIFFERENTIAL EQUATION

Consider an elementary control volume with co-ordinated system x-streamwise parallel to the bed, y-lateral, and z-normal to the bed. Assuming steady uniform flow, the Navier-Stokes equation for a fluid element may be written in the x-direction as

$$\rho g \sin\theta + \frac{\partial \tau_{yx}}{\partial y} + \frac{\partial \tau_{zx}}{\partial z} = \rho\left(\overline{V}\frac{\partial \overline{U}}{\partial y} + \overline{W}\frac{\partial \overline{U}}{\partial z}\right) \tag{1}$$

where $\overline{U}$, $\overline{V}$, $\overline{W}$ are the local mean velocity components in the x, y, z direction; ρ is the density of the water; g is the gravitational acceleration; θ is the channel bed angle; τ_{ij} is the shear stress in the j-direction on the plane perpendicular to the i-direction. The depth-averaged momentum equation may be obtained by integrating Eq. (1) over the local water depth H. This procedure is described elsewhere by Shiono & Knight (1988) and Shiono & Knight (1991), where it is shown that Eq. (1) becomes

$$\rho g H S_0 - \rho\frac{f}{8}U_d^2\sqrt{\left(1+\frac{1}{s^2}\right)} + \frac{\partial}{\partial y}\left(\rho\lambda H^2\sqrt{\frac{f}{8}}\,U_d\frac{\partial U_d}{\partial y}\right) = \frac{\partial}{\partial y}\left[H\left(\rho\overline{UV}\right)_d\right] \tag{2}$$

where U_d is the depth mean velocity integrated between $z_b = z_b(y)$ (local bed elevation) and z_s (water level), that is $U_d = \dfrac{1}{H}\displaystyle\int_0^H \overline{U}\,dz$ (3)

where $H(y) = z_s - z_b(y)$ as illustrated in Fig. 1.

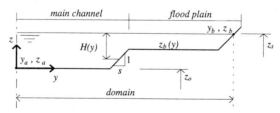

FIGURE 1 • Notation for the geometric characteristics across the section of the channel

On the LHS of Eq. (2), S_0 is the channel bed slope; s is the channel side slope (1:s, vertical:horizontal); f is the local friction factor; and λ is the dimensionless eddy viscosity. The term on the RHS of Eq. (2) represents the depth averaged effects of secondary flow, also denoted by Γ. Equation (2) is therefore the governing differential equation which describes the lateral distribution of depth mean velocity in a prismatic channel flowing in a steady state condition.

FINITE ELEMENT APPROACH

Assuming that all the coefficients in the governing equation (2) are solely dependent on the lateral distance y, and expressing the channel side slope in general terms as $1/s = \partial z_b(y) / \partial y$, the following substitutions may be introduced

$$U(y) = U_d(y)^2 \qquad\qquad \alpha(y) = \frac{\rho}{2}\lambda(y)H(y)^2\sqrt{\frac{f(y)}{8}}$$

$$\beta(y) = -\rho\frac{f(y)}{8}\sqrt{1+\left(\frac{\partial z_b(y)}{\partial y}\right)^2} \qquad q(y) = \frac{\partial}{\partial y}\left[H(y)\left(\rho\overline{UV}\right)_d\right] - \rho g H(y) S_0 \qquad (4)$$

Equation (2) is thus reduced to

$$\frac{\partial}{\partial y}\left(\alpha(y)\frac{\partial U(y)}{\partial y}\right) + \beta(y)U(y) = q(y) \qquad (5)$$

Therefore the governing equation has now been simplified into linear second order differential equation in which the independent variable is the velocity squared, represented by U. The boundary conditions (BC) and the domain for this equation are

Domain: $\qquad\qquad\qquad\qquad y_a < y < y_b \qquad\qquad\qquad\qquad\qquad (6a)$

Boundary conditions: $\quad$ at $y=y_a \qquad U(y_a) = U_a \quad$ or $\quad \left(\dfrac{\partial U}{\partial y}\right)_{y_a} = 0 \qquad (6b)$

$$\text{at } y=y_b \qquad U(y_b) = U_b \quad \text{or} \quad \left(\frac{\partial U}{\partial y}\right)_{y_b} = 0 \qquad (6c)$$

In order to apply the Finite Element Method, the next step is the discretisation of the domain into elements interconnected with each other at common points called nodes. The independent variable U is then substituted by the trial or approximate solution u within each element *(e)* in the form: $U^{(e)}(y) \approx u^{(e)}(y; a) = \sum\limits_{j=0}^{n} a_j N_j^{(e)}(y) \qquad (7)$

where $N_j^{(e)}$ are the shape functions; and a are the unknown parameters located at the nodes of the mesh. Applying Galerkin Weighted-Residual as the method to minimise the error introduced by the trial solutions [see Burnett (1986)], Eq. (2) may be transformed into a system of simultaneous equations of the form $[K]\{a\}=\{F\}$, where

$$K_{ij}^{(e)} = \int_{(e)} \frac{\partial N_i^{(e)}}{\partial y}\alpha\frac{\partial N_j^{(e)}}{\partial y}dy - \int_{(e)} N_i^{(e)}\beta\, N_j^{(e)}\, dy \qquad F_i^{(e)} = -\int_{(e)} N_i^{(e)} q\, dy \qquad (8)$$

Once the shape functions are selected according to the type of elements, the contributions of each element are assembled into the global system of equations. Ultimately the boundary conditions expressed by Eqs. (6) will then define the entire problem. The resulting set of values a, called degrees of freedom DOF, represents the approximate solution to the problem. Important physical quantities can thus be

3

derived from the solution, namely boundary shear stress τ_b, and depth averaged Reynolds shear stress $\bar{\tau}_{yx}$, as

$$\tau_b = \frac{f}{8}\rho U_d^2 \qquad \text{(9a)}$$

$$\bar{\tau}_{yx} = \rho\lambda\,H\frac{1}{2}\sqrt{\frac{f}{8}}\,\frac{\partial U_d^2}{\partial y} \qquad \text{(9b)}$$

APPLICATION OF THE NUMERICAL MODEL

A FE-based computer model called RFMFEM has been implemented in order to perform a numerical analysis of the problem. The computer program was developed in C++ language and allows the combination of linear and quadratic elements in the same mesh. A complete description of the model is given by Abril (1995).

TRAPEZOIDAL CHANNEL (neglecting secondary flow contribution $\Gamma=0$)
Benchmarking and validation of the model are achieved by comparing the performance of both types of elements against the analytical solution [Shiono & Knight (1991)] for the trapezoidal channel in Fig. 2 with the following characteristics

Geometric data: $\quad H=1$ m, $b=1$ m, $s=1$, $S_0=0.001$

Hydraulic properties are constant: $\quad f=0.08$, $\lambda=1.0$, $g = 10$ m s^{-2}, $\rho=1000$ kg m^{-3}

Boundary conditions: $\quad$ *Essential BC:* $\quad U_d=0 \quad$ at $y=2.0$ m

$\qquad\qquad\qquad\qquad$ *Natural BC:* $\quad \partial U_d/\partial y = 0 \quad$ at $y=0.0$ m

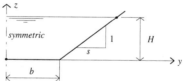

FIGURE 2 • Geometric description of the semi-domain of a trapezoidal channel

In order to test the convergence of the method, several simulations were carried out increasing the number of degrees of freedom (DOF) for 2-node linear and 3-node quadratic elements. The Finite Element solution for U_d are compared with the analytical solution in Table 1.

	Linear Elements			Quadratic Elements			Analytical	% Error 17DOF	
y	5DOF	9DOF	17DOF	5DOF	9DOF	17DOF	solution	Linear	Quadratic
0.00	0.7882	0.8177	0.8281	0.8089	0.8245	0.8304	0.8337	-0.67	-0.40
0.50	0.7821	0.8126	0.8233	0.8035	0.8197	0.8257	0.8291	-0.70	-0.41
1.00	0.7633	0.7968	0.8086	0.7868	0.8046	0.8111	0.8149	-0.77	-0.47
1.50	0.7173	0.7566	0.7712	0.7285	0.7655	0.7742	0.7793	-1.04	-0.65
1.75	-------	0.6983	0.7204	-------	0.6976	0.7250	0.7337	-1.81	-1.19

TABLE 1 • Finite Element solution for U_d compared with the analytical solution

Clearly, the error increases towards the discontinuity produced at $y=2.0$ m. Therefore, the accuracy can be improved by using the combination of a coarse mesh of quadratic elements with a refined discretisation of linear elements close to this region. The FE solutions for velocity U_d and Reynolds shear stress $\bar{\tau}_{yx}$ with this hypothesis compared against the analytical solution are illustrated in Fig 3.

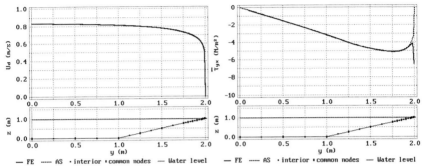

FIGURE 3 • Comparison of Finite Element solution for U_d and $\bar{\tau}_{yx}$ against the analytical solution

COMPOUND CHANNEL (SERC-FCF)

For the case of a compound channel the geometric description is taken from the Science and Engineering Research Council-Flood Channel Facility series of experiments 020501 for which the general characteristics are:

Geometric data: $H=0.19796$ m, $s=1$, $S_0=0.001027$

Constant hydraulic properties: $g = 9.807$ m s^{-2}, $\rho=1000$ kg m^{-3}

The local hydraulic properties are assumed to be constant within the main channel and flood plain regions and to have a linear variation on the sloping side of the main channel as

Flood plain properties: $f=0.0163$, $\lambda=0.070$, $\Gamma=1.4667$

Main channel properties: $f=0.0239$, $\lambda=0.633$, $\Gamma=-2.376$

The results obtained by the model for velocity U_d and boundary shear stress τ_b are compared with the corresponding experimental data, as illustrated in Fig. 4. Further details on the calibration philosophy are given by Knight & Abril (1995).

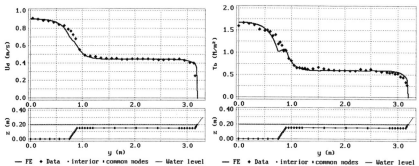

FIGURE 4 • Comparison of Finite Element solutions for U_d and τ_b against SERC-FCF 020501 data

CONCLUSIONS

A depth-averaged FE-based numerical model for the description of the lateral distribution of velocity, Reynolds and boundary shear stresses in compound channels has been introduced. A comparative analysis and validation of the numerical solution against the analytical solution proved the model to be highly accurate even with the use of a coarse mesh such as the one shown in Fig. 3. The model presented a remarkably good agreement when compared with actual data especially regarding boundary shear stress distribution. This is a promising feature if further sediment transport analysis is to be included, though a careful consideration must be undertaken with respect to the calibration philosophy.

REFERENCES

1. Abril, J.B., 1995, "Numerical modelling of turbulent flow and sediment transport by the finite element method", MPhil Thesis, The University of Birmingham, England, UK.

2. Burnett, D.S., 1987, "Finite element analysis, from concepts to applications", Ed. Addison-Wesley by AT&T Bell laboratories, USA.

3. Knight, D.W., and Abril, J.B., 1995, "Refined calibration of a depth averaged model for turbulent flow in a compound channel", *Hydra 2000, Proc. 26th IAHR Congress*, London, UK, Sep, pp?-?.

4. Knight, D.W., and Shiono, K., 1995, "River channel and flood plain hydraulics", in Flood plain processes [Eds. M.Anderson, P.Bates and D. Walling], Chapter 5, J. Wiley & Sons.

5. Shiono, K., and Knight, D.W., 1988, "Two-dimensional analytical solution for a compound channel", *Proc. 3rd Int. Symposium on Refined Flow Modelling and Turbulence Measurements*, Tokyo, Japan, July, pp. 503-510.

6. Shiono, K. and Knight, D.W., 1991, "Turbulent open channel flows with variable depth across the channel", *Journal of Fluid Mechanics*, Vol. 222, pp.617-646 (and Vol.231, October, p. 693).

EXPLICIT FORMULAE FOR ESTIMATION OF NORMAL DEPTH IN CHANNELS

J. ATTARI, A.M. FOROGHI
M.Sc. Students, Tarbiat Modaress University, Tehran, Iran

ABSRACT
In open channel flow, computation of uniform flow depth often involves trial and error. Some explicit formulae for estimation of this depth for rectangular, trapezoidal and circular cross sections have been developed by using curvefitting methods. Comprison of the results show that the formulae estimate the normal depth with acceptable accuracy for practical works.

INTRODUCTION
Computation of uniform flow depth (normal depth) is a basic problem in open channel hydraulics. For this purpose usually Manning´s equation (1) is applied.

$$Q = \frac{A}{n} R^{2/3} S^{1/2} \qquad (1)$$

For known values of discharge (Q), roughness coefficient (n), longitudinal slope (S), cross sectional area (A) and hydraulic raduis (R), solving equation (1) often implies trial and error or numerical solution. The dimensionless form of the Manning´s equation(2), is suitable for graphical solution.

$$\frac{nQ}{b^{8/3} \sqrt{S}} = \frac{AR^{2/3}}{b^{8/3}} = K \qquad (2)$$

Where b is the bottom width of the section and for circular section it will be replaced by d, which is the diameter. The dimesionless factor K represents the conveyance of the channel. For any cross section, K is only a function of the dimensioless term (y/b) :

$$K = f\left(\frac{y}{b}\right) \qquad (3)$$

Since the values of Q, n, S, b are usually known, K can be determined from equation (2). On this basis, tables of K versus (y/b) have been prepared which are useful for determination of normal depth (y_n).

METHOD OF SOLUTION

In order to find explicit formulae, equation (3) should be rearranged in the following form :

$$\frac{y}{b} = f_1 (K) \tag{4}$$

To solve this problem, a curve was fitted to generated data to find an explicit expression for y/b such as equation (4). This technique has been applied to channels with trapezoidal, rectangular and circular cross sections but the procedure is only demonstrated for trapezoidal section in detail here.

FORMULA FOR TRAPEZOIDAL SECTION

Substituting the geometrical characterstics of a trapezoidal section in equation (3) yields:

$$K = \frac{(1 + Z \, y/b \,)^{5/3}}{[1+2 \, y/b \, (1+Z^2)^{1/2}]^{2/3}} \, (\frac{y}{b})^{5/3} \tag{5}$$

Data was generated by assuming arbitrary values for the variables of right hand side of equation (5) and computing corresponding values for K. Graphs of such data on a log-log scale have been plotted years ago by Chow [1]. These curves appear to be straight lines, with the correlation coefficient of 0.98 . It was intended in this study to fit a single line to the data instead of a family of curves with higher accuracy. In the first step, the following model was adopted for the curve fitting:

$$\frac{y}{b} = a_1 K + a_2 K^m \tag{6}$$

This model can be transformed to the following linear model:

$$\log (\frac{y}{bK} - a_1) = \log a_2 + (m-1) \log K \tag{7}$$

Assuming $U = \log (\frac{y}{bK} - a_1)$ and $V = \log K$, eq.(7) presents a straigth line:

$$U = a_3 + a_4 V \tag{8}$$

In equation (7) only a_2 and m was computed by the least square method, and a_1 was determined by trial and error to obtain the best correlation factor. Equation (8), represents a straight line for each value of sidewall

8

slope of trapezodial section Z. By changing Z, a family of straight lines were derived. In the second step, the purpose was to represent these lines by a single equation involving Z. Assuming a semi-log correlation between a_4 and Z results:

$$a_4 = a_5 + a_6 \log Z \tag{9}$$

No	Formula & Range of Applicablity	K	Section
1	$y_n = b\ (1.155\ K + 2/3\ K^{0.547})$ $0.02 < y_n/b < 1.5$	$\dfrac{nQ}{b^{8/3}\sqrt{S}}$	Rectangle
2	$y_n = b[(1-0.05\ L_n\ Z\ )\ K^{0.6} - 0.2\ K^{1.3}\ ]$ $1 < Z < 2$ $0.03 < y_n/b < 0.3$	$\dfrac{nQ}{b^{8/3}\sqrt{S}}$	Trapezoid
3	$y_n = b\ [(1.145 - 0.4 \log Z)\ K^{0.61} - K/3.5]$ $1 < Z < 2$ $0.3 < y_n/b < 1$	$\dfrac{nQ}{b^{8/3}\sqrt{S}}$	Trapezoid
4	$y_n = \dfrac{K^{3/8}\ [4\ (1+Z^2)\ ]^{1/8}}{Z^{5/8}}$	$\dfrac{nQ}{\sqrt{S}}$	Triangle
5	$y_n = d\ (0.9\ K + 0.823\ K^{0.433}\ )$ $0.04 < y_n\ /d\ < 0.77$	$\dfrac{nQ}{d^{8/3}\sqrt{S}}$	Circle

Table 1   Explicit formulae for estimation of normal depth

From equations (8) and (9) the sidewall slope Z appears in the equation:

$$U = a_3 + (a_5 + a_6 \log Z) V \qquad (10)$$

Finally, by using a computer program a two independent variable correlation was computed. By substituting the computed coefficients and after simplification, final formulae were obtained. The new correlation coefficient for trapezoidal section was found to be 0.9995. In order to have higher accuracy, two diffrent equations for the trapezoidal section and their corresponding ranges of applicablity are given in table 1.

FORMULAE FOR OTHER SECTIONS
A similar procedure has been adopted for the other sections,except for triangular section which has an implicit equation.The formula for estimation of the normal depth y_n are given in table 1.

ACCURACY OF THE FORMULAE
The comparison between exact computed values with the results of formulae, are summarized in table 2 for different sections. As it can be obsereved the relative error for rectangular, trapezodial and circular sections are 1.4%, 3.4% and 3.9% respectively.

APPLICATION
To show the application of formulae three examples for common cross sections have been selected from [2], and the summary of the results are given in table 3.

| Cross | Given Data | | | | | y_n (m) | | Formula |
| Section | b or d | S | n | Q | Z | Comp. | Formula | No. |
	(m)	x10^4		(m^3/s)		Value	Value	
Rectangle	6	197	0.02	11	0	1.0	0.992	1
Trapozeid	3	16	0.13	7.1	1.5	2.6051	2.6058	2
Circle	0.91	16	0.015	0.42	-	0.535	0.541	5

Table 3 Summary Of The Examples

Z	K	Yn/b(exact)	Yn/b(form)	%Er
	0.0029	0.030	0.0298	0.63
	0.0160	0.084	0.0827	1.56
1	0.0367	0.138	0.1349	2.23
	0.0641	0.192	0.1868	2.72
	0.0980	0.246	0.2384	3.10
	0.1382	0.300	0.2898	3.41
	0.0029	0.030	0.0294	2.08
	0.0165	0.084	0.0824	1.90
1.5	0.0385	0.138	0.1359	1.49
	0.0686	0.192	0.1902	0.96
	0.1068	0.246	0.2450	0.40
	0.1532	0.300	0.3004	0.14
	0.0029	0.030	0.0291	3.04
	0.0169	0.084	0.0824	1.93
2	0.0402	0.138	0.1372	0.59
	0.0727	0.192	0.1936	0.82
	0.1147	0.246	0.2513	2.17
	0.1669	0.300	0.3102	3.40

(a) Trapezoid
(formula NO. 2)

Z	K	Yn/b(exact)	Yn/b(form)	%Er
	0.1382	0.30	0.3029	0.97
	0.2727	0.44	0.4403	0.08
1	0.4526	0.58	0.5766	0.58
	0.6811	0.72	0.7112	1.22
	0.9615	0.86	0.8432	1.95
	1.2973	1.00	0.9714	2.86
	0.1532	0.30	0.2984	0.54
	0.3144	0.44	0.4407	0.15
1.5	0.5395	0.58	0.5834	0.58
	0.8352	0.72	0.7241	0.57
	1.2078	0.86	0.8606	0.07
	1.6636	1.00	0.9905	0.95
	0.1669	0.30	0.2960	1.32
	0.3530	0.44	0.4420	0.45
2	0.6205	0.58	0.5885	1.47
	0.9795	0.72	0.7319	1.65
	1.4394	0.86	0.8682	0.96
	2.0095	1.00	0.9942	0.58

(b) Trapezoid
(formula NO. 3)

K	Yn/b(exact)	Yn/b(form)	%Er
0.0014	0.02	0.0202	1.10
0.0671	0.23	0.2292	0.33
0.1671	0.44	0.4427	0.62
0.2799	0.65	0.6541	0.64
0.3991	0.86	0.8624	0.28
0.522	1.07	1.0675	0.23
0.6472	1.28	1.2698	0.80
0.78	1.50	1.4790	1.40

(c) Rectangle
(formula NO. 1)

K	Yn/d(exact)	Yn/d(form)	%Er
0.0008	0.04	0.0390	2.43
0.0073	0.11	0.1046	1.22
0.0279	0.20	0.1999	1.16
0.0692	0.32	0.3212	0.31
0.1304	0.45	0.4581	1.53
0.2017	0.59	0.5931	1.26
0.2677	0.71	0.7060	1.11
0.2996	0.79	0.7600	3.86

(d) Circle
(formula NO. 4)

TABLE 2 COMPARISON OF RESULTS

11

CONCLUSION

The tables of comparison of the results show that for ordinary engineering work, application of these simple formlae have enough accuracy. Having explicit relation between y_n/b and K might be useful when combination of this equation with other governing equations is desired.

ACKNOWLEGEMENTS

The authors would like to express their appreciation to Dr. H. Tadayon for his supervision of the study, and the Water Research Center in Tehran for their kind cooperation.

NOTATIONS

a_i	numerical coefficients
A	flow cross- sectional area
b	bottom width of the channel
d	diameter of circle
f, f_1	function
K	conveyance of the channel
log	logarithm in the base 10
L_n	logarithm in the base e
n	Manning´s roughness coefficient
m	numerical coefficient
Q	discharge
R	hydraulic radius
S	longitudinal slope of the channel
U,V	variables
y	depth of flow
y_n	uniform flow depth (normal depth)
Z	side wall slope of the channel (Horiz : 1)

REFERENCES

[1] Chow, V.T., "Open Channel Hydraulics", McGraw Hill Book Co., Int. Student edition, Tokyo, 1959, p 130

[2] French, R.H. "Open - Channel Hydraulics", McGraw Hill Book Co., Int. Student edition, Singapore, 1986, pp 166,172

ABOUT THE DECOMPOSITION OF WAVES BY A SUBMERGED HORIZONTAL PLATE

MICHAEL BECKER
Student, University of Wuppertal
Pauluskirchstr. 7, 42285 Wuppertal, Germany

The physical processes near a horizontal submerged plate applied beneath waves are very complicated. The process of particular interest for this work was the decomposition of waves passing the plate. It was found that the transformation of waves has no influence on the transmitted energy and that no transformation of wave frequencies occurs in the range of wave length where the largest flow beneath the plate was noticed.

REASONS FOR THE INVESTIGATION

A horizontal submerged plate used for reducing the wave height has several advantages compared to conventional breakwaters. Its operation as a wave filter allows a profitable solution which makes also possible an ecologically desirable exchange of water between the open sea and the area to be protected.

A critical comparison of the literature [Graw, 1994 and 1993] shows many phenomena, but not all of them can be related with each other. It appeared that by varying all parameters the hydrodynamical system close to the plate reacted very sensitively upon changes, so that an enormous program of measurements was necessary for an analytical description.

The comparison of the literature with respect to the decomposition of waves above the plate shows that very different results were published. Many authors do not remark on this topic; one reason may be that they did not notice any decomposition. Dattari et al. [1977] remark that they did not observed a decomposition; Dick [1968] noticed a production of harmonic waves as well as Guevel et al. [1985] and Kojima et al. [1990] did. Guevel et al. [1985] explained the observation with eddy separation at the plate creating a second order wave. Kojima et al. [1990 and 1992] gave detailed analysis of the magnitude of the higher order waves, they explained the influence of frequency superposition of the incoming waves and described a numerical model for the calculation of the decomposition. But even with these publications the knowledge about this phenomenon is still limited. Therefore detailed experiments were made, which are described here.

Furthermore, there is another important reason for this kind of analysis: the submerged horizontal plate changes the spectral energy distribution behind the building. Even if this would not have a substantial influence on the total energy contained in the different waves of the spectrum, it has still to be regarded as a change. Because of the suitability of the submerged horizontal plate especially for the protection of ecologically sensitive areas the knowledge about the frequency change is of extraordinary importance for the planning of the breakwater.

EXPERIMENTS

The variation of the wave length was limited by the wave generator and the plate length was modified according to potential constructions. It was necessary to vary the parameter L/l gradually with very small changes because an extra-ordinary discontinuous behaviour had to be described. This was achieved by varying the wave length L, the sub-mergence depth d_p and the wave height H_i for plates of different lengths.

18 series of measurements were ana-lysed: plate lengths of $l = 0.50$ m, 0.75 m and 1.00 m, for each of them at a sub-mergence depth of $d_p = 0.02$ m, 0.03 m and 0.04 m for an initial wave height of $H_i = 0.02$ m and $d_p = 0.03$ m, 0.045 m and 0.06 m for $H_i = 0.03$ m.

ANALYSIS

First of all, all results of the measure-ments were shown graphically with respect to the wave length L of the initial wave. The transmitted wave height H_t, the mean value (v_m) and the fluctuation amplitude (v_a) of the velocity beneath the plate were plotted. The figures 1 to 3 show three typical examples.

By means of a Fourier's sequence it is possible to examine periodical functions concerning their frequencies. First of all

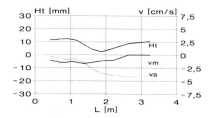

Fig. 1: v, H_t-values at the 0.50 m-plate ($d_p = 0.04$ m, $H_i = 0.02$ m)

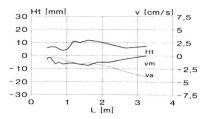

Fig. 2: v, H_t-values at the 0.75 m-plate ($d_p = 0.03$ m, $H_i = 0.02$ m)

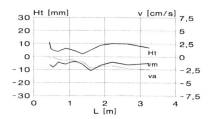

Fig. 3: v, H_t-values at the 1.00 m-plate ($d_p = 0.03$ m, $H_i = 0.02$ m)

the results of the evaluation program showing the percentage of the transmitted wave heights at the corresponding frequency of the initial wave height were documented graphically.

Figure 4 shows an exemplary diagram: the shares of the fundamental oscillation and the different harmonic oscillations are visible. Figure 5, however, shows an example where the harmonic oscillation is not a multiple of the fundamental oscillation. For a plate with $l = 1.00$ m, $d_p = 0.04$ m and $H_i = 0.02$ m the 1st harmonic oscillation of the fundamental frequency $f = 1.2$ Hz is not (approx.) 2.4 Hz, but between 2.4 Hz and 2.7 Hz. In the following this case is called leakage as common in this kind of analysis.

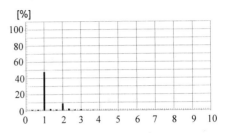

Fig. 4: Result of a Fourier's analysis

The results of one series of measurements (i. e. for a plate length l, submergence depth d_p and an initial wave height H_i) were summarised in one diagram afterwards (figs. 6 to 7 show the diagrams corresponding to figures 1 to 3), whereas the frequency shares are represented in form of curve diagrams. This implies a kind of filter effect because the single values are not exactly connected. The graphs representing the mean values facilitate the analysis of general trends.

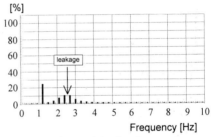

Fig. 5: Leakage of the first harmonic

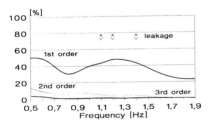

Fig. 6: Frequency distribution:
$l = 0.50$ m
$(d_p = 0.02$ m, $H_i = 0.02$ m)

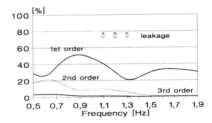

Fig. 7: Frequency distribution:
$l = 0.75$ m
$(d_p = 0.03$ m, $H_i = 0.02$ m)

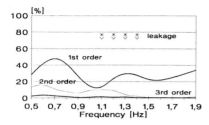

Fig. 8: Frequency distribution: $l = 1.00$ m
($d_p = 0.03$ m, $H_i = 0.02$ m)

EVALUATION OF THE FREQUENCY ANALYSIS FOR EACH SINGLE PLATE

On closer examination of the frequency analysis for each single plate one can see that the course of the fundamental oscillations is similar to the course of the C_t-graphs. In most cases the fundamental oscillation comprises the largest share of the wave height, the 1st harmonic oscillation the second largest. The configurations represented in table 1 can be regarded as an exception of this general behaviour.

Plate length l	Submergence depth	1st harmonic oscillation nearly as large as the fundamental oscillation at:	1st harmonic oscillation larger than the fundamental oscillation at:	Minimum C_t-value at:
[m]	[m]	[Hz]	[Hz]	[Hz]
$H_i = 0.02$ m				
0.50	0.03	0.8	-	0.8
	0.04	-	0.7 - 0.8	0.8
0.75	0.03	-	0.6	0.6
	0.04	-	0.5 - 0.6	0.6
1.00	0.02	-	1.0	0.9 - 1.0
	0.03	1.0 - 1.1	-	1.0
	0.04	1.1	0.5	1.1
$H_i = 0.03$ m				
0.50	0.045	0.8	-	0.8
0.75	0.045	-	0.7	0.7
	0.06	-	0.7	0.8
1.00	0.06	0.7	-	0.7

Table 1: Configurations with harmonic oscillation ≥ fundamental oscillation

The wave frequencies for which a 1st and/or 2nd harmonic oscillation being larger than the fundamental frequency was observed are frequencies near the frequency of the best wave height reduction (first C_t-minimum). It is probable that this minimum is caused by a conversion of the initial wave into a double frequency wave, this means the energy is withdrawn from the general system by means of this conversion.

Small frequencies create harmonic oscillations, the larger do not. Harmonic waves with double or three times the fundamental frequency occur only in this case, they can be ignored regarding large frequencies.

In the area of frequencies with the largest currents observed no conversion of wave length can be noticed. It is probable that the energy necessary for the production of higher harmonics near the C_t-minimum mentioned is converted in a current flow around the plate in the case of larger frequencies.

A leakage of the results for the medium frequencies (1.0 to 1.4 Hz) concerning the 1st harmonic oscillation can be noticed.

It is striking that the largest shares in the harmonic oscillation can be seen at 0.6/0.7 Hz. There is a second minimum at 1.1/1.2 Hz. An analysis effect (choice of the evaluation intervals, digitalisation frequencies) and a channel effect (2nd order waves are formed during wave production) are excluded because it is not possible to prove this behaviour by means of the analysis of the initial wave.

An examination of all the figures of C_t-coefficients shows that the intensity of the higher frequencies (visible for the 2nd order) is growing depending on increasing submergence depth. The harmonic conversion is the first effect mentioned here which is influenced by the arrangement of the plates. The increase of the leakage also seems to correspond to the submergence depth.

In any case a phenomenon was detected here that in contrary to the present points of interest does not depend on the relationship wave length to plate length but on the period of the wave! An examination whether the water depth is the decisive parameter which determines the process is necessary.

It is also necessary to qualify the suggestions made according to which the reduction of the wave height can be attributed to the conversion into waves of higher order. The reduction by the horizontal submerged plate breakwater only seems to affect the 1st order wave. If the 1st order is severely reduced and the generation of harmonics is strong at the same time finally the 2nd or higher order wave will be obviously larger than the fundamental oscillation.

CONCLUSIONS

The results supplement the existing knowledge in the following subjects:

♦   The transformation of the initial wave into waves of higher order by the break-water has no influence on the amount of energy transmitted, nevertheless it might be important for the ecological situation of the area protected by the breakwater.

♦   No transformation of wave frequencies occurs in the range of wave length where the largest flow was noticed.

♦   There are considerable indications that the transformation of the initial wave into waves of higher order depends on the water depth. Further investigations are necessary. It could be shown that the transformation neither depends on the plate length nor on the immersion depth.

OUTLOOK

The analysis of the decomposition of waves by the submerged horizontal plate presented here was only a small part of a larger research project. The main object of the investigation was to examine the physical phenomenon of the pulsating flow beneath the plate. In order to understand all conditions influencing this phenomenon a comprehensive variation of parameters is necessary, entailing an immense number of experiments. Several dependencies could be demonstrated by the investigations but up to now it was not possible to give an analytical description.

REFERENCES

Becker, M.; *Labormessungen und Energiebetrachtungen zum Zusammenhang von Wirkungsgrad des Plattenwellenbrechers und Plattenumströmung;* 1993; Diplom-arbeit, Wasserbau und Wasserwirtschaft, Bergische Universität Wuppertal

Dattari, J.; Jothi Shankar, J.; Raman, H.; *Laboratory investigations of submerged platform breakwaters;* 1977; Proc.: 17th Int. IAHR Congress, Karlsruhe, pp.89-96

Dick, T.M.; *On solid and permeable submerged breakwaters;* 1968; Ph. D. Diss. at Queens University Kingston

Graw, K.-U; *Untersuchungen am Plattenwellenbrecher;* 1994; Mitteilung Nr. 7, Lehr- und Forschungsgebiet Wasserbau und Wasserwirtschaft, Bergische Universität Wuppertal

Graw, K.-U.; *The submerged plate as a primary wave breaker;* 1993; Proc.: XXV IAHR Congress, Tokyo, pp.38-45

Graw, K.-U.; *The submerged plate as a wave filter: the stability of the pulsating flow phenomenon;* 1992; Proc.: 23rd ICCE, Venice, pp.1153-1160

Kojima, H.; Yoshida A.; Ijima, T.; *Second-order interactions between waves and a submerged horizontal plate;* 1992; 23rd ICCE, Venice, pp.241-242 (abstracts)

Kojima, H.; Ijima, T.; Yoshida, A.; *Decomposition and interception of long waves by a submerged horizontal plate;* 1990; 22nd ICCE, Delft, pp.1228-1241

The horseshoe vortex in super-critical flow

A.H.BENSON
University of Nottingham, England

SUMMARY

The flow field around the base of a circular cylinder in super-critical flow was investigated experimentally. Dye flow visualisation, velocity measurements and pressure measurements in the plane of symmetry confirmed the existence of a horseshoe vortex at high Froude numbers. The influence of Froude number on the flow field was studied.

INTRODUCTION

It is well-known that a circular cylinder mounted vertically on a flat surface induces a three-dimensional separation of the boundary layer, which curls over to form a horseshoe vortex. This vortex can act as a scour agent, and is of particular interest for it's influence on scour around bridge piers situated in rivers. Much work has been carried out on the nature of the vortex in air (Baker, 1980) and in sub-critical water flow conditions (Dargahi, 1989), but little has been done under super-critical flow conditions. White(1975) and Jain and Fischer(1980) studied scour for Froude numbers up to 1.5, but did not look at the vortex itself. Ettema(1980) stated that a horseshoe vortex did not form on the planar bed around a pier if the approach flow was super-critical.

This paper outlines an experimental investigation into the nature of the flow around a pier mounted in super-critical flow. Dye flow visualisation results are given, together with velocity profiles and pressure distributions along the line of symmetry on the approach to the pier.

EXPERIMENTAL RIG

The experiments were carried out in a 4.7 metre long, 90 cm wide and 42 cm deep, glass-sided recirculating flume, with adjustable slope and discharge. A circular cylindrical perspex tube of diameter 6 cm was attached to the base of the flume 3.7 metres from the inlet. For the purposes of flow visualisation and pressure measurement, tappings were sunk into the base of channel at 0.3 cm intervals from 0.2 cm to 2 cm upstream of the pier, and at 1 cm intervals from 2 cm to 10 cm upstream of the pier. Polythene tubes were sealed onto the lower ends of the tapping tubes, through which potassium permanganate solution could be injected. The flow pattern was photographed from the side of the flume.

Velocities within the flow were measured using a miniature conical hot-film probe mounted on the overhead instrument carriage. This was used with an anemometer, the output of which was taken into a computer and sampled using a data acquisition package. Data was analysed using Microsoft Excel. The differential pressure between a reference tapping 10 cm upstream of the pier and the tappings closer to the pier was measured by means of a differential pressure transducer, with a range of 0-10 mBar. Tube lengths between tapping and transducer were minimised to increase frequency response, and the electrical signal produced by the transducer was again taken to the computer for sampling and analysis.

Seven flow conditions were used throughout the work, with Froude number as the main variable. This was varied from 1.46 to 2.54 by altering the slope of the flume, and hence the flow velocity, while maintaining a constant flow depth of 3 cm.

RESULTS

It can be seen from the flow visualisation pictures of Figure 1 that a vortex is present at the leading edge of the pier. Another noticeable flow feature is the large bow wave which increases in height with Froude number. The flow curls up the face and sides of the pier, falling back on itself in the upstream direction. The influence of the bow wave extends up to 5 cm upstream.

The velocity profiles measured in the open channel (Figure 2a) follow the log-law as expected, and collapse onto each other well when non-dimensionalized with the mean velocity, with the exception of Run 1. The turbulence intensity plot (Figure 2c) shows turbulence intensity to be greater near the bed, as expected. Turbulence intensity decreases with Froude number.

The velocity profiles measured in the separation zone (Figure 2b) are similarly sensible, taking a slight S-shape and tending towards zero near the bed as the flow meets the reverse flow of the vortex. The profiles get steeper as Froude number increases, but mean velocity occurs at the same height for all flow conditions, at about 1.3 cm above the bed. The turbulence intensities (Figure 2d) are much larger in the separation zone than in the open channel, possibly due to the lower mean velocity used in calculation. The spectra taken with the hot-film (Figures 2e & 2f) show a broad spread of energy across the frequency range.

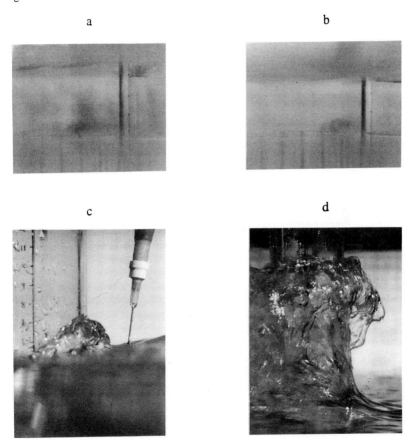

Figure 1:   Flow visualisation results
   a -   Vortex for F=1.46   c -   Bow wave for F=1.46
   b -   Vortex for F=2.42   d -   Bow wave for F=2.42

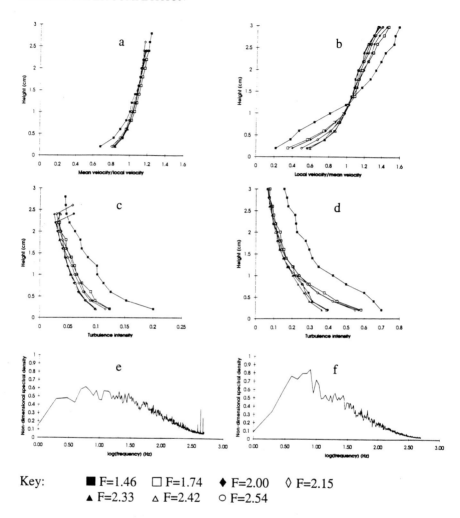

Key: ■ F=1.46   □ F=1.74   ♦ F=2.00   ◊ F=2.15
        ▲ F=2.33   △ F=2.42   ○ F=2.54

Figure 2:   Velocity measurements along plane of symmetry, upstream of pier
    a -   Velocity profile in open channel (x=10 cm)
    b -   Velocity profile at separation zone
    c -   Turbulence intensity profile in open channel (x=10 cm)
    d -   Turbulence intensity profile at separation zone
    e -   Power spectrum in open channel (x=10 cm), 0.5 cm above bed
    f -   Power spectrum at separation zone, 0.5 cm above bed

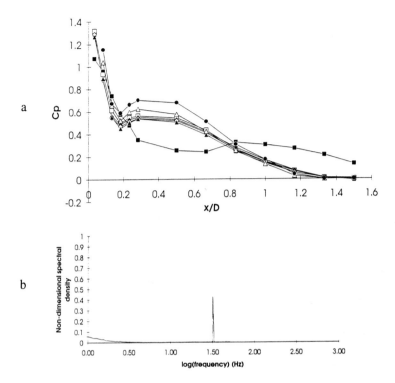

Figure 3:   Pressure measurements upstream of pier, at bed level
 a -   Pressure coefficient profile along plane of symmetry
 b -   Power spectrum of pressure variation

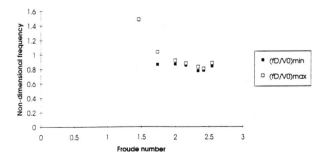

Figure 4:   Variation of pressure fluctuation frequency with Froude number

From Figure 3a, it can be seen that the pier causes an increase in pressure on the upstream bed, the influence of the pier extending upstream by at least 1.333 diameters for all Froude numbers. In the case of Run 1, however, the influence extends much further upstream, perhaps due to the slower velocity near the bed. A local minimum in pressure coefficient can be seen near the pier, which corresponds with the position of the vortex centre. With the exception of Run 1, this occurs consistently at x/D=0.183. The pressure then rises steeply towards the pier, the maximum pressure here being due to either the bowwave or downflow. The results are in agreement with those of Baker(1980).

DISCUSSION AND CONCLUSIONS
In conclusion, a horseshoe vortex exists upstream of a circular cylinder mounted in super-critical flow, and the influence of the pier extends over one pier diameter upstream, even in high Froude number flows.

ACKNOWLEDGEMENTS
The author wishes to acknowledge the help of her supervisor, Dr.C.J.Baker, throughout the course of the work, and Mr.M.Langford for taking photographs.

NOTATION

$C_p = p - p_0 / \frac{1}{2} \rho V^2$	Pressure coefficient
D	Pier diameter (cm)
f	Frequency (Hz)
$F = V/\sqrt{gh}$	Froude number
h	Flow depth (cm)
p/p_0	Pressure/Pressure at reference point (N/m^2)
V/V_0	Velocity/Mean velocity (m/s)
x	Distance upstream of pier leading edge (cm)

REFERENCES
Baker,C.J.(1980), "The turbulent horseshoe vortex", J.Wind Eng.&Ind.Aero., 6, pp.9-23.

Dargahi,B.(1989), "The turbulent flow field around a circular cylinder", Exp.in Fluids, 8, pp.1-12.

Ettema,R.(1980), "Scour at bridge piers", School of Engineering Report 216, University of Auckland, Auckland, New Zealand.

Jain,S.C. & Fischer,E.E.(1980), "Scour around bridge piers at high flow velocities", J.Hyd.Div., Proc.ASCE, 106(HY11), pp.1827-1843.

White,W.R. (1975), "Scour around bridge piers in steep streams", Proc.16thIAHR Congress, Sao Paulo, 2, pp.279-284.

Data Assimilation and Parameter Estimation in a 2-D Advection-Dispersion Model

RAFAEL CAÑIZARES TORRE
Hydroinformatics Research Fellow (ICCH - IHE)
International Centre for Computational Hydrodynamics,
Hørsholm, Denmark
Institute for Infrastructure, Hydraulics and Environmental Engineering,
Delft, The Netherlands

ABSTRACT

An extended Kalman filter for the assimilation of observed data and parameter estimation, has been incorporated into the advection-dispersion module of a standard 2D hydrodynamic modeling tool. The Kalman filter is applied only to the dispersion part to ensure a constant influence of the filter, so avoiding expensive computational effort. Equations have been added to estimate unknown parameters in order to obtain a more accurate model. The filter performance has been tested and some application results are presented.

INTRODUCTION

New measurement technologies have emerged for the observation of natural processes that occur within many of the different scientific fields where numerical models are used. Large amounts of data are now available to test and to improve the accuracy of these models, and this in turn generates a need to assimilate the measured data efficiently into the model.

Most data-assimilation applications are in the field of meteorology, but recently these have also appeared in oceanography and hydrology. Stochastic processes play a very important role when the models are used for forecasting. Therefore, techniques such as adjoint methods and Kalman filters have been applied to these fields, albeit with varying degrees of success. The number of data assimilation applications in the field of hydrodynamics is still quite sparse.

DATA ASSIMILATION

The aim of data assimilation is to improve the solution of the numerical models, generally closing the gap between model solution and observations in a least-squares sense. Three different data assimilation techniques have been applied in many scientific fields. *Optimal interpolation* has been used primarily in meteorological numerical weather prediction. Optimal interpolation treats the noise in the observations as part of the observation and assume a time independent error. *Variational data assimilation* has been applied in oceanographic models. In this technique the numerical model and its adjoint are integrated together to obtain the optimal solution. *Kalman filter* applications can be found especially in studies of groundwater and hydrology. The Kalman filter is similar to the optimal interpolation method but as in variational data assimilation it has the advantage of taking into account the data noise effects in the estimation of the model variables, so providing an estimate of the final model error. This last feature makes the two latter methods computationally expensive.

A distinction must be made between data assimilation and parameter estimation. Data assimilation corrects the model output without any correction in the state parameters. Parameter estimation corrects the model parameters while the model output is not corrected directly. A combination of both techniques improves considerably the accuracy of model results.

KALMAN FILTER

The Kalman filter is the statistically optimal method for sequential assimilation of data into linear numerical models. The original filter (Kalman 1960; Kalman-Bucy, 1961) was applied to linear equations. The filter can also be applied to non-linear systems, expanding the non-linear equations about the current best estimate of the actual state. This method is called the Extended Kalman filter. The Kalman filter, which is based on the Gauss Least Squares Method, provides an estimate of the state of the system at the current time based on all measurements of the system obtained up to and including the current time. The description of the model dynamics is explicitly introduced into the Kalman filter. Therefore the estimation of non-measured state variables is allowed because the model dynamics relate the state and the measured variables in a consistent way.

The extended Kalman filter formulation follows two steps: the forecast and the measurement update. In the first step the new state of the system is determined using the model equations. In the second step, measured data are combined with the forecast state and the weighting between the measured value and the forecast is used to improve the estimate of the model state.

The magnitude of model and measurement errors are assumed known. It is assumed that these errors have Gaussian distribution with zero mean. In addition, initial estimations of error covariances, representing the uncertainty of the initial conditions, are needed to initiate the filter algorithm.

In the Kalman filter, the error covariance function (covariance of the difference between model and data) is used to compute a correction to the model. The estimate of the covariance is updated after every time step, and it is this which makes the Kalman filter computationally very expensive.

APPLICATION TO A 2-D ADVECTION-DISPERSION MODEL
In the present application two objectives have been considered. The first is assimilation of measured data into the model with the aim of correcting the model output and estimating unknown physical parameters. The second objective is to reduce the high computational effort associated with the Kalman filter.

In this specific application, the mass transport equation is divided into two parts: a dispersion and an advection part. The advection equation has time dependent coefficients (dependent on the velocity field) unlike the dispersion equation if the diffusion coefficient remains invariant. Using superscripts D and A for dispersion and advection, respectively, we get the model equation in the following form;

$$X^{t,D+A}_{estimate} = \Phi X^{t-1,D+A}_{analysis} + \Phi_A X^{t-1,D+A}_{analysis}$$

The vector X contains all model variables. The matrix Φ represents the dispersion equation coefficients while the matrix Φ_A contains the advection equation coefficients. The applied filter consists of the following algorithm:

$$X^{t,D}_{estimate} = \Phi X^{t-1,D+A}_{analysis}$$
$$P^t_{estimate} = \Phi P^{t-1}_{analysis} \Phi^T + Q$$
$$K^t = P^t_{estimate} H^T (HP^t_{estimate} H^T + R)^{-1}$$
$$P^t_{analysis} = (I - K^t H^T) P^t_{estimate}$$
$$X^{t,D}_{measured} = \Phi X^{t,D+A}_{measured} + \Phi_A X^{t-1,D+A}_{analysis}$$
$$X^{t,D}_{analysis} = X^{t,D}_{estimate} + K^t (X^{t,D}_{measured} - H X^{t,D}_{estimate})$$
$$X^{t,D+A}_{analysis} = X^{t,D}_{analysis} + \Phi_A X^{t-1,D+A}_{analysis}$$

Where P, Q and R are the error covariance matrices of the difference between model and data, the model error and the measurement error respectively, H is the observation matrix and K is the Kalman gain matrix or weight matrix. The observational data is influenced by advection and diffusion, but only the diffusion part is used in the filter. This is done by subtracting the best estimation of the advection part up to the current time step.

The standard deviation of system errors, measurement errors and errors in the initial condition are usually unknown. Consequently, the filter must be calibrated and the correlation between the errors at different grid points and the magnitudes of the variances must be estimated based on the initial values of the problem. These magnitudes are critical when some of the parameters, which can affect the numerical stability of the model, are estimated.

PARAMETER ESTIMATION
The procedure used for parameter estimation follows Moll & Crebas, 1988. The parameter to be estimated is included in the state vector by means of a scalar, f_D, which is used to multiply the parameter in the state equation. A new equation which relates f_D at two adjacent time steps, is included into matrix Φ.

The update in f_D is proportional to the errors in the measurement points and to the factors included in the corresponding row of the Kalman gain. These values depend on the error covariance at the measurement points, on measurement errors, and on correlations between errors in the parameter and model errors. These correlations are usually unknown and must be previously estimated because they are the most influential factors in the parameter estimation.

MASS CONSERVATION CONSTRAINT
An additional constraint can be introduced into the filter in order to obtain more accurate estimates. This constraint is conservation of mass and is included as an additional equation in the matrix Φ. The total mass is conserved, allowing for losses or gains due to outflows and inflows at every time step.

APPLICATION CASE
In order to validate the filter two tests cases have been built. The first tests the filter under an advective situation. The longitudinal velocity is constant and equal to 0.4 m/s and the transversal velocity is zero. A reference model run is made to generate the "measurements". This model use a diffusion coefficient of $1 m^2/s$. A second model has been run using "measurements" at seven positions and the Kalman filter, and an initial diffusion coefficient of $0.4 m^2/s$. The filter uses parameter estimation and mass conservation. Model parameters are: $\Delta x = \Delta y = 10m$; $\Delta t = 5$ sec. After 100 time steps, the diffusion coefficient estimated was $0.65 m^2/s$ and the results are shown in figure 1.

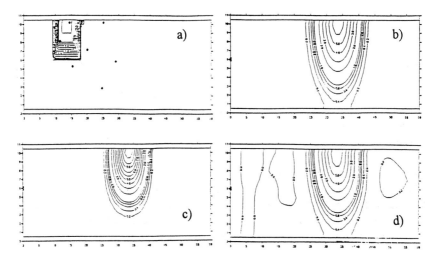

figure 1. Advection test. a) initial conditions and measurements location; b) "measurements" (reference model result); c) result when D = 0.4m²/s is used without data assimilation d) result when D = 0.4m²/s is used with data assimilation.

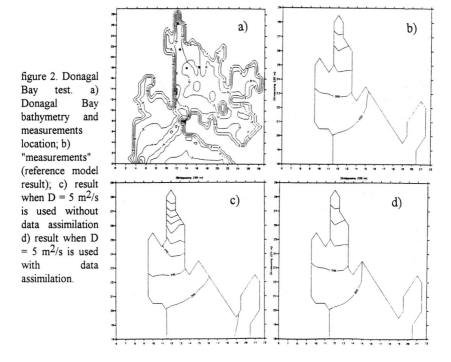

figure 2. Donagal Bay test. a) Donagal Bay bathymetry and measurements location; b) "measurements" (reference model result); c) result when D = 5 m²/s is used without data assimilation d) result when D = 5 m²/s is used with data assimilation.

Another more realistic test case is a model of Donagal Bay. The same procedure to create "measurements" have been used but with a diffusion coefficient of $D = 20 \ m^2/s$. Four measurement positions are used. As an initial value for the assimilation test, we use $D = 5 \ m^2/s$. Model parameters are : $\Delta x = \Delta y = 220m$; $\Delta t = 120$ sec; and the model is run for 400 time steps. A constant discharge of 340 units/s acts as a concentration source in the bottom of the bay. Advection in the model is caused by semi-diurnal tidal wave with 1 m. amplitude. The diffusion coefficient estimated was $19.8 \ m^2/s$ using the filter, and the results are shown in figure 2.

CONCLUSION

A Kalman filter has been applied into the 2-D advection-dispersion module in the model MIKE 21. The filter shows very promising results in the two test cases performed as data assimilation combined with parameter estimation tool.

ACKNOWLEDGMENTS

This works forms part of ongoing post graduate research carried out under an IHE, Delft - ICCH / DHI, Denmark, joint collaboration. The author wishes to thank these Institutes for providing the facilities and sponsorship. I thank my supervisors Prof. M.B. Abbott, T.G. Jensen and H.J. Vested for their support and guidance.

REFERENCES

1. GHIL M. & MALANOTTE-RIZZOLI P. 1991, *Data Assimilation in Meteorology and Oceanography,* Advances in Geophysics Volume 33
2. HEEMINK A.W. 1988 , *Two-dimensional shallow water flow identification.* Appl. Math. Modeling, 1988, Vol.12, April
3. KALMAN R.E. 1960, *A new approach to linear filtering and prediction problems.* Journal of Basic Engineering, March 1960 , 35-45
4. LONG R.B 1988, *Notes on assimilating observations into numerical models.* Delft Hydraulics.
5. MOLL J.R. & CREBAS J.I. 1989, *An operational management system for river flows,* Delft Hydraulics Publication, number 423
6. SMEDSTAD O.M. & FOX D.M. 1994, *Assimilation of Altimeter Data in a Two-Layer Primitive Equation Model of the Gulf Stream,* American Meteorological society
7. THACKER W.C. 1988, *Three lectures on fitting numerical models to observations.* GKSS 87/E/65
8. ZHOU Y. 1991, *Kalmod, A stochastic-Deterministic Model for Simulating Groundwater Flow with Kalman Filtering.* IHE report series 22

Influence of Spatial Variability of Hydraulic Parameters on Aerobic Biotransformations in Groundwater Systems

OLAF CIRPKA
Institute of Hydraulic Engineering, University of Stuttgart, Germany

ABSTRACT

Results of model calculations are presented that demonstrate how spatial variability of hydraulic aquifer properties is influencing microbial activity. Concentration profiles along streamlines are similiar to profiles in one-dimensional models, wheras concentration profiles integrated over the width of the two-dimensional domain indicate higher mixing. As longitudinal mixing processes dominates over transversal mixing a model of non-interacting stream pipes is suggested as alternative to two-dimensional stochastical modelling.

INTRODUCTION

During the last decade *in-situ* stimulation of microbial activity has turned out as economical approach for cleaning up aquifers contaminated by well degradable organic compounds. In most applications electron acceptors like oxygen or nitrate and, if necessary, additional nutrients are injected into the groundwater by injection wells. Efficiency of the method is dependent on the delivery of those compounds to microbes present at the location of contamination. Hence groundwater flow and advective-dispersive transport are key processes to be evaluated. Especially the high spatial variability of hyraulic conductivity in the subsurface with variations in the order magnitudes is leading to a limitation of electron acceptor and nutrient delivery.

GOVERNING EQUATIONS

For each dissolved compound i being in equilibrium to sorption the transport equation in form of (1) is valid:

$$n_e R_i \frac{\partial c_i}{\partial t} + \mathrm{div}\left(n_e \underline{v}_e c_i - n_e \underline{\underline{D}}\,\mathrm{grad}(c_i)\right) = n_e r_i \qquad (1)$$

R_i nominates the retardation factor. The sink-source term r_i includes chemical transformations. Assuming a simple microbial system with the components oxygen c_O, substrate c_S and biomass X the growth rate might be expressed by a double Monod expression (2):

$$k_{gr} = \mu_{max} \frac{c_O}{c_O + K_O} \frac{c_S}{c_S + K_S} \tag{2}$$

The reactions rates of oxygen r_O and substrate r_S are proportional to the biomass growth (3-4):

$$r_O = \frac{1}{Y_O} k_{gr} X \tag{3}$$

$$r_S = \frac{1}{Y_S} k_{gr} X \tag{4}$$

For development of biomass an additional linear decay rate k_{dec} is assumed (5):

$$\frac{\partial X}{\partial t} = \left(k_{gr} - k_{dec}\right) X \tag{5}$$

Equations (1) to (5) have to be solved for the simulation of aerobic biodegration in the subsurface. Transport of microbes might be introduced by an additional transport equation. Kinetic sorption would have to be expressed by additional ordinary differential equations instead of the retardation factor. The stiff non-linear equation system is solved by an iterative coupling scheme, in which advective-dispersive transport is calculated for each mobile compound by Finite Element modelling and microbial processes are calculated nodally by Gear's stiff method [Cirpka&Helmig 1994].

NUMERICAL MODEL RESULTS

ONE-DIMENSIONAL TEST CASE
In order to illustrate the dynamic behaviour of the system a simple one-dimensional test problem has been chosen. Assume a homogeneous soil colum 100m in length contaminated by a sorbing contaminant. No oxygen is present. Initial biomass concentrations are very low. From timepoint zero on oxygen enriched water is injected continuously into the colum. No substrate is present in the injected water. Parameters of the test case are shown in table 1. Length profiles of calculated concentrations are shown in figure 1.

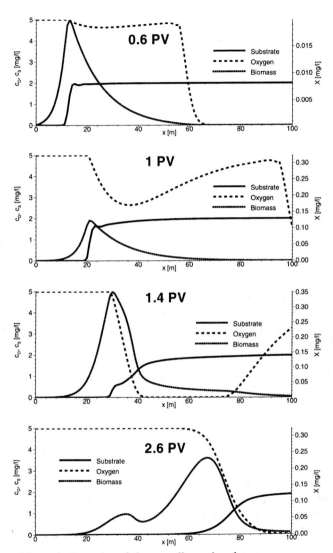

Fig. 1   Numerical results of the one-dimensional test case.

Table 1   Parameters of the one-dimensional test case.

· Flow and Transport Parameters				
n_e=0.3	R_S=5	K=10^{-3}m/s	$\partial h/\partial x$=10^{-2}	α=10^{-2}m
· Microbiological Parameters				
μ_{max}=0.5/d Y_S=.09		Y_O=.032	K_S=2.0mg/l K_O=0.2mg/l k_{dec}=0.05/d	
· Initial Conditions		c_S=2mg/l c_O=0mg/l X=10^{-3}mg/l		
· Inflow Concentrations		c_S=0mg/l c_O=5mg/l		
· Outflow Boundary Condition		Pure Advection		

As oxygen is assumed not to be retarded, the front velocity of the injected oxygen is higher than the front velocity of the washed out, sorbing contaminant (chromatographic effect). As a consequence in the occuring mixing zone both oxygen and substrate are available to the microbes. They start to grow exponentially. Initially the biomass concentration is very low and hence the reaction rates for oxygen and substrate are hardly influencing transport. With growing biomass the importance of reactive processes is increasing. After a certain time the total mass flux of oxygen is consumed by microbial processes. The system has changed from advection dominated to reaction dominated.

The remaining mixing zone is quite small. Inside of this zone biomass concentrations reach their maximum. The zone is moving with a characteristic velocity somewhere between the effective velocity of oxygen and of substrate in the non-reactive case.

HETEROGENEOUS TWO-DIMENSIONAL TEST CASE
The two-dimensional system is a confined aquifer 100m in length and 20m in width. The geometric mean of hydraulic conductivity is 10^{-3}m/s, standard deviation of logarithmic hydraulic conductivity σ_{lnK} is 1.0. The distribution was generated by a sequential Gaussian simulator [Deutsch&Journel, 1992]. Correlation length are 5m in length direction and 2.5m perpendicular to it. Hydraulic head is fixed at the left and right boundary (head difference 1m), the upper and lower boundaries are impermeable. Longitudinal dispersivity is 0.1m and transversal dispersivity 0.01m. Microbial parameters, initial and boundary conditions are the same as in the 1D test case. Concentration distributions for the timepoint 1.5 pore volumes after begin of injection are shown in figure 2.

It is obvious that dissolved compounds are transported faster in higher permeable zones. This is leading to heterogeneous concentration patterns.

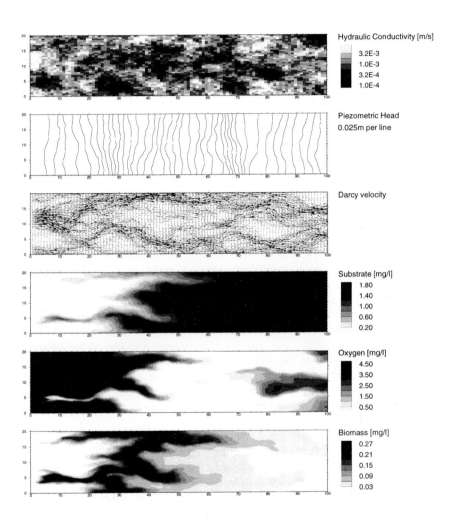

Fig. 2   Numerical results of the two-dimensional heterogeneous test case.
Timepoint 1.5 PV after begin of oxygen injection.

Preferential places for biomass growth are downstream interfaces of low permeable zones. Here substrate mass flux out of the low permeable zone is remaining for a long time, while delivery of oxygen is ensured due to transport through the high permeable zone. Notice that to a certain - not identified - extend mixing is an artefact of finite element modelling.

DISCUSSION AND CONCLUSIONS

Integration of concentrations over the width of the domain would lead to length profiles indicating high longitudinal mixing. This macrodispersion would be observed as well in breakthrough curves of extraction wells. It exceeds local mixing by far. As microbial processes are taking place on the pore scale macrodispersion is not relevant for the estimation of mixing in the context of advective-dispersive transport coupled to biotransformations. One dimensional modelling with macrodispersion as dispersion coefficient leads to an overestimation of local mixing and hence of microbial activity.

In contrast to that concentration profiles along streamlines are quite similiar to the 1D test case. As effective velocity along different streamlines differs, concentration profiles along different streamlines differ as well. Interaction between those 'slow' and 'fast' streamlines is not very strong, since longitudinal dispersion exceeds transversal dispersion. Additional mixing processes like chromatographic effects and kinetical desorption act as well as longitudinal 'dispersion'. Hence in a simplified approach transversal mixing might be neglected.

This is leading to a model of non-interacting streampipes with varying hydraulic conductivities. A set of 1d models differing in hydraulic conductivity has to be calculated independently. For achieving integrated profiles or breakthrough curves the results are to be weighted by the probability density function of arrival times. In contrast to stochastical 2D or 3D modelling this approach gives the opportunity to calculate complex reactive interactions in heterogeneous media with a reasonable computational efford. Comparative calculations are on the way.

REFERENCES

Cirpka, O., R.Helmig: Numerical Simulation of Contaminant Transport and Biodegradation on Porous and Fractured-Porous Media, in A. Peters et al. (eds.): *Computational Methods in Water Resources X*, pp.605-612, Kluwer Academic Publishers, Dordrecht, 1994.

Deutsch, C.V., A.G. Journel: *GSLIB-Geostatistical Software Library and User's Guide*, Oxford Univesity Press, New York, 1992.

DUNE VELOCITY IN SAND BED RIVERS

JUAN JOSE FEDELE
Student of Water Resources Engineering
Faculty of Engineering and Water Sciences
Littoral National University
Santa Fe - Argentina

INTRODUCTION

The reliable field measurement of bed forms displacement velocity, u_b, in alluvial streams, has been always a considerable difficult and/or expensive task. As a consequence many mathematical formulas and methods have been developed in order to predict u_b (see for example, Graf, 1971 and Yalin, 1977).

The gross of that formulas have two basic drawbacks at the moment of their use in practice: a) their application range is limited, and b) the parameters on which they are based, are not easy to estimate in field.

In this paper, the calibration of a semiempirical formula that enables the prediction of u_b in sand water streams of a wide range of sizes: from laboratory channels to large plain rivers, is presented. The field measurement of their parameters with the standard hydrometric techniques, is very simple.

THEORY

It is generally accepted that the gross of bed load formulas can be reduced into a simple expression of the following type:

$$g_{sf} = k (\tau_*' - \tau_{*c})^p \qquad (1)$$

where: g_{sf}, bed load transport; τ_*', dimensionless grain shear stress; τ_{*c} ,dimensionless shear stress for initiation of motion; k, p, constants to be determined.

Expressing the effective dimensionless shear stress, $\tau_*'- \tau_{*c}$, as a function of the total dimensionless shear stress, τ_* (through, for example, the Engelund's (1982) relation) and introducing the Manning equation:

$$g_{sf} = k \left[\frac{0.3}{(s-1)^{3/2}} \right]^p \left[\frac{(u \cdot n)^3}{d^{3/2} \cdot h^{1/2}} \right]^p \qquad (2)$$

where: u, mean velocity of flow; n, Manning roughness coefficient; s, specific gravity of bed sediment; d, representative diameter of bed sediment; h, flow depth.

As it is well known in sand bed streams with bed forms:

$$g_{sf} = \alpha (1-P) H u_b \qquad (3)$$

where: α, dune shape coefficient ($\approx \frac{1}{2}$ for triangular dunes); P, bed material porosity; H, mean height of bed forms.

Equating (2) and (3) and solving for u_b, it is obtained:

$$u_b = \frac{k}{\alpha \cdot (1-P) \cdot H} \left[\frac{0.3}{(s-1)^{3/2}} \right]^P \left[\frac{(u \cdot n)^3}{d^{3/2} \cdot h^{1/2}} \right]^P \qquad (4)$$

It is seen in eq. (4) that if n is replaced by a predictor e? pressed as a function of the same variables of (4), i.e.: d, H and h, it would be possible to have an equation for u_b depending only on that three variables and u, all of them easily measurable in field.

DEVELOPMENT OF THE n PREDICTOR

According to Yalin (1992), the Chezy's dimensionless friction factor, c, of an undulated surface in a rough turbulent flow, can be expressed generally as:

$$c = \Phi_c (h/k_s ; H/\lambda ; \lambda/h) \qquad (5)$$

where: k_s, granular roughness of bed surface, and λ, bed form lenght.

Introducing in (5) the relationship between c and n, and taking into account that $k_s \propto d$ and $\lambda \propto h$, it results:

$$n = \Phi_n (h/d ; H/h) \cdot h^{1/6} \qquad (6)$$

with:

$$\Phi_n = c'_g [\Phi_c']^{-1} \quad \text{and} \quad c'_g = g^{-1/2}$$

The form of eq.(6) was set empirically through the following procedure based on selected series of the laboratory data reported by Guy et al. (1966):

a) It was assumed that n is the sum of the two classic components: n' (grain roughness) and n" (form resistence). The n' values were predicted through a Manning-Strickler type equation fitted to the plain and transition bed data. It was obtained a very good fit ($r^2 = 0.77$) with the following equation:

$$n' = 0.057 \cdot d^{1/6} \qquad (7)$$

b) The observed values of n" were computed substracting $\acute{n}'$ to the observed values of n. Several trials were then carried out for each considered bed diameter in order to adjust a relationship between n" and H/h^k (k = real number). The best fits were obtained with k = 1/3. In this way, a family of rigth lines n" = m (H/h$^{1/3}$) passing through the origin and with increasing m as a function of d, was defined. All those right lines were represented by the following single equation:

$$n" = 1.504 \; d^{0.39} \; H/h^{1/3} \qquad (8)$$

Finally, adding (7) and (8):

$$n = h^{1/6} \; [\; 0.057 \; (d/h)^{1/6} + 1.504 \; d^{0.39} \; H/h^{1/2}] \qquad (9)$$

In Fig. 1 are shown the n obs values vs. the n values computed with eq. (9) and the scatter bands of ± 20 %.

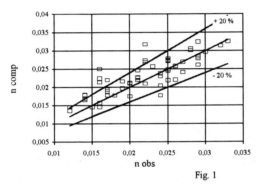

Fig. 1

Introducing (9) in (4), rearranging and simplifying, it results the following equation for u_b:

$$u_b = c'_{ub} \; \frac{1}{H} \left[\frac{1}{d} \cdot (0.057 + 1.504 \cdot d^{2/9} \cdot \frac{H}{h^{1/3}})^3 \right]^p \left[\frac{u^3}{h^{1/2}} \right]^p \qquad (10)$$

with:

$$c'_{ub} = \frac{k}{\alpha \cdot (1-P)} \left[\frac{0.3}{(s-1)^{3/2}} \right]^p$$

Eq. (10) is dimensionally consistent provided that the adequate dimensions in the constants of eqs. (1), (7) and (8), be considered.

CALIBRATION

Three sources of dunes velocity data were used to calibrate eq. (10): the sets of laboratory data reported by Guy et al. (1966); records corresponding to large and small superimposed dunes of the Paraná river (FICH, 1992; Amsler and Gaudin, 1994) and of the Paraguay river (HRS, 1972). The u, h, H , λ and d values ranges of the field measurements were the following:

	Paraná river		Paraguay river
u [m/s]	0.70 - 1.80		0.75 - 1.02
h [m]	7 - 16		5 - 11.9
d [mm]	0.2 - 0.4		0.24
H [m]	1 - 4	(large dunes)	0.90 - 1.20
	0.15 - 0.60	(small dunes)	
λ [m]	50 - 200	(large dunes)	33 - 49.50
	5 - 10	(small dunes)	

Eq. (10), was set into the dimensionless form: $u_b \, H / \sqrt{g \cdot d^3}$ before the calibration. It is seen that this parameter would be nothing else but the dimensionless volumetric bed load transport per unit width due to the dunes displacement. Scatter graphs of the data points showed two arrangements of the points: one corresponding to the smallest diameters ($d < \sim 0.4$ mm) and the other to the larger ones. Thus, two equations were fitted to each group. The r^2 values were not sensitive to the different p values. In fact, with p varying between 0.8 and 1.8, it was obtained a $r^2 \cong 0.77$ for $d < \sim 0.4$ mm and a $r^2 \cong 0.95$ for the larger diameters. Thus it was selected the classic $p = 1.5$ that yielded the two following equations (Figs. 2 and 3):

$d < \sim 0,4$ mm

$$\frac{u_b \cdot H}{\sqrt{g \cdot d^3}} = 5.7 \cdot 10^{-9} \left[(1 + 26.4 \cdot d^{\cdot 22} \cdot \frac{H}{h^{1/3}})^{3.56} \cdot \frac{u^{3.56}}{d^{2.4} \cdot h^{0.6}} \right] \qquad (11)$$

$d > \sim 0,4$ mm

$$\frac{u_b \cdot H}{\sqrt{g \cdot d^3}} = 1.5 \cdot 10^{-9} \left[(1 + 26.4 \cdot d^{\cdot 22} \cdot \frac{H}{h^{1/3}})^{4.05} \cdot \frac{u^{4.05}}{d^{2.7} \cdot h^{0.68}} \right] \qquad (12)$$

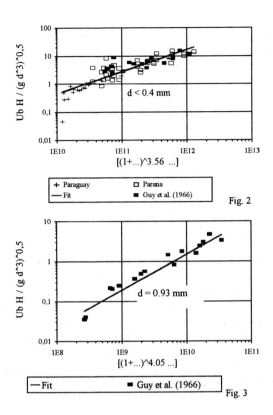

Fig. 2

Fig. 3

In Fig. 4 the values of u_{bo} vs u_{bc} with the scatter bands of ±50 % are presented.

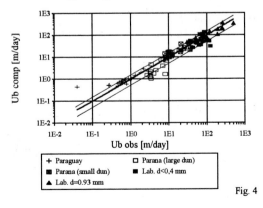

Fig. 4

CONCLUSIONS

• A semiempirical equation based on parameters easily measurable in field that enables the prediction of dunes velocity in alluvial streams of a wide range of sizes, was obtained. Its constants and exponents vary for bed particles diameters smaller or larger than ~ 0.4 mm.

• The arrangement of the data points depending on the bed sediment size would be not surprising since, with the smaller diameters, the viscous effects not considered in the analysis, are not negligible. It would be also reflected in the different values obtained for r^2.

• In the derivation of eq. (10), it was necessary to develop a resistence predictor based on the classical division for grain and form friction and adjusted with laboratory data.

• It must be pointed out that for the computation of the field data resistence during the calibration, the small superimposed dunes height was used. It agrees with the observation of Znamenskaya (1963) and the results of Amsler and Schreider (1992) for the Paraná river, with regard to the outstanding role played by the small superimposed dunes in the form resistence carried by natural streams. This aspect is discussed with greater detail in a forthcoming paper together with a verification of eqs. (11) and (12) based on data sets different from those of the calibration.

REFERENCES

AMSLER, M.L. and GAUDIN, H.E., 1994, "Influence of the superimposition of dunes on the bed load transport in the Paraná river", XV Water National Congress, La Plata, Argentina (in spanish).

AMSLER, M.L. and SCHREIDER, M.I., 1992, "Hydraulic features of the superimposition of bed forms in the Paraná river (Argentina)", Proceedings of the XV Latinamerican Congress of Hydraulics, IAHR, Vol. 3, Cartagena, Colombia, p.39-48 (in spanish).

ENGELUND, F. and FREDSØE, J., 1982, "Sediment ripples and dunes", Ann. Rev. Fluid Mechanics, Vol. 14, p.13-37.

FICH (FACULTY OF ENGINEERING AND WATER SCIENCES), 1992, "Design of the dredged area overlenght for the subfluvial tunnel protection building", FICH - Interstate Manager Committee of the Subfluvial Tunnel Uranga-S. Begnis, Progress Report, July 1992 (in spanish).

GRAF, W.H., 1971, "Hydraulics of Sediment Transport", Mc Graw-Hill Book Co. Inc., New York.

GUY, H.P.; SIMONS, D.B. and RICHARDSON, E.V., 1966, "Summary of alluvial channel data from flume experiments, 1956-61", Prof. Paper 462-I, U.S. Geological Survey.

HRS (HYDRAULICS RESEARCH STATION), 1972, "River Paraguay Study", Report No. EX 606, Wallingford, Berkshire, England.

YALIN, M.S., 1977, "Mechanics of Sediment Transport", Pergamon Press, 2nd Edition, Oxford, England.

YALIN, M.S., 1992, "River Mechanics", Pergamon Press, Oxford, England.

ZNAMENSKAYA, N.S., 1963, "Experimental study of the dune movement of sediment", Transaction of the State Hydrologic Institute (Trudy GGI), No 108 (translated by L.G. Robbins).

3-D wave overtopping on caisson breakwaters

CLAUDIO FRANCO
University of Rome, Italy

ABSTRACT
A 3-D model investigation has been carried out within a joint European research project, in order to assess the effects of the obliquity and "multi-directionality" of wave attack on the overtopping performance of different types of caisson breakwaters. The parametric influences on wave overtopping of structure geometrical changes and wave conditions are described in terms of simple correction factors to be applied to the abundant results obtained with the simple vertical face structure under long-crested normal attack. A general expression is derived and compared with the overtopping performance of sloping structures.

INTRODUCTION
In the past, short-crested waves have been reproduced in physical models only for rather complex cases, such as the definition of loads on off-shore platforms or the correct reproduction of the vessels movements inside harbours .Recently, 3-D models have also been used for reasearch on coastal structures and beaches, such as studies on wave overtopping and run-up on slopes (de Waal and V.d. Meer, 1992) and the response of a sandy beach (Franco C. et al., 1994) . Very little attention has been addressed so far to the tridimensional response of vertical structures, and in particular to the overtopping performance .

Wave overtopping over vertical-face caissons is indeed, one of the most relevant hydraulic responses, since it affects the safety of transit or mooring on the breakwater crown, which is quite common in Mediterranean harbours .

The main reference for the design of vertical structures against wave overtopping is Goda (1985), based on early 2-D tests on a simple vertical wall. A new comprehensive 2-D model study on the overtopping of typical deepwater caisson breakwaters has been presented by Franco L. (1993,1994). After the latter work, the present 3-D "extension" has been designed, and the author took charge of the data analysis after direct partecipation in the entire test programme .

MODEL SETUP AND WAVE CONDITIONS
Physical model tests were carried out in the "Vinjè basin" at Delft Hydraulics laboratory in the Netherlands, during summer 1994 . The structure (Fig. 1),

consisting of 13 caissons, each 0.9m wide, was placed on a flat, 0.61m deep, concrete bottom, and linked to the floor through a permeable two layer rock basement .

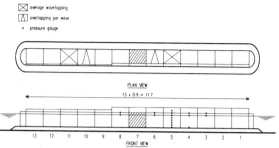

Figure 1 : Front view of the 13 caissons with indication on the placement of the overtopping measuring devices

In order to reduce the number of test to be performed the structure was instrumented in two parts each with different crest height . Overtopping measuring devices consisting of 1.0m wide water tray intake, were placed along each side for direct evaluation of the average discharges. Crest elements were made removeable in order to adjust the relative crest height or freeboards R_c/H_s of 1.18 ; 1.5 and 1.63 .

The target significant wave height was kept fixed to 0.14m throughout all tests that were run with no less than 1000 waves. Peak periods (Tp) of standard Jonswap spectra were 1.5 s and 2.12 s to give theoretical peak wave steepnesses $s_{op}=2\pi H_{os}/gT_p^2$ of 0.04 and 0.02 respectively. Wave attack angles varied from normal (0°) to the most oblique (60°) referred to the perpendicular to the structure. Energy dispersion around the mean direction was Gaussian with standard deviations σ set equal to 0°, 15° and 20°.

An extensive set of tests were performed for the simplest case of straight vertical plain wall, whereas a more limited number of tests were conducted for the alternative

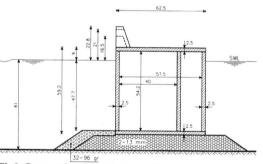

Fig.2: Cross-section of the caisson with the simple vertical wall

structure geometries . In one alternative configuration a small additional recurved nose was placed at the top of the parapet wall. Two new structural layouts were tested after placing a 1:3 impermeable smooth concrete slope connected to the caisson frontal wall just below the water level, either directly or through a 1m wide berm . Three other configurations were modelled to observe the response of perforated walls with circular (ϕ =51mm) and rectangular holes (36x72mm). The perforated walls had, overall wall porosity equal to 20%, overall wet perimeter constant and for rectangular holes, a height to base ratio equal to 2 . At last, together with the rectangular perforated wall, an air intake has been opened in the caisson upper deck to simulate the effect of air vents that are typical in many prototype caissons A total of 84 tests with non-breaking wave conditions (a part from the test with slopes and with berms in front of the structure) were performed

ANALYSIS OF MEAN OVERTOPPING DISCHARGES

The basic assumption, confirmed by many previous researchers (Owen,1980; V.d.Meer,1994; Franco,1993), is that the main parameters influencing the overtopping average discharge q, i.e. the crest height R_c, wave height H_s are related through adimensional numbers by the exponential law :

$$Q = a \exp (-bR) \qquad (1)$$

where Q is the adimensional discharge defined as $Q= q / (gH_s^3)^{1/2}$, R is the relative freeboard $R=R_c/H_s$ and a and b are coefficients which depend on other factors except wave height and crest heights. For the simple case of plain vertical wall, with long-crested waves normal to the structure Franco L.(1994) gathering many European model test results, found a=0.19 and b=4.2 . Goda (1985) presented a set of design curves that can be translated into equation (1) with a=0.045, nearly constant, and b variable depending on the relative depth h_t/H_s , on the wave steepness and on the sea-bed slope . The data points obtained from this 3-D study under the same unidirectional wave attacks compare quite well with the "two-dimensional" prediction line from Franco (1993), even though points show some scatter, while the curves from Goda (1985) give an underestimation of about half order of magnitude on the overtopping rates . Based on the assumption that the influences of freeboard and wave height follow the relationship (1), the influences of wave obliquity, multidirectionality and steepness can be assessed by means of a set of reduction factors γ to be included into (1) in the form :

$$Q = a \exp (-bR/ \gamma) \qquad (2)$$

where $\gamma =b_{ref.}/b_{fitting}$ and a=0.19 is set constant. $b_{ref.}$ is the angular coefficient of the reference case that is according to Franco L. (1993) equal to 4.2, and $b_{fitting}$ is the angular coefficient of the regression lines that describe the overtopping response under different wave conditions . Therefore the value of γ directly represents the potential reduction of freeboard to accomodate the same overtopping rate under similar incident wave heights . Even though only

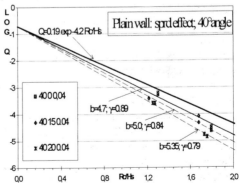

Fig.3 : Example of data fitting . Different degrees of spreadings with fixed 40° attack angle .

few data points for each wave condition were available, the example of data regression reported in figure 3 shows that the fitting was generally good. For each caisson geometry, γ values have been plotted against β , the angle of attack. Each data set refers to a wave condition with a different degree of energy spreading (0°, 15° and 20°) and/or different wave steepness. In figure 4, which is referred to the plain wall case, it can be seen that :

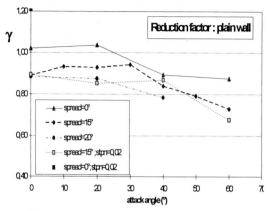

Fig 4: Reduction factors. Plain vertical wall

- Overtopping reduction due to wave obliquity is not defined in the range 0° to 30°, being nearly constant, whereas from 30° to 60° the γ drops to 0.9 for long-crested waves (σ =0°) and to 0.75 for short-crested waves .
- The multidirectionality effect is difficult to assess, but generally it is given in a 10% reduction of γ when referred to the long-crested case . Very little difference is found between the two degrees of spreading (15°- 20°) . The influence of wave steepness is not apparent for short-crested waves (σ>0°) .

In general, similar conclusions as for the plain wall case, could be derived for the other caissons configurations. The above statements suggested to consider separately only the two cases of short-crested and long-crested waves for which regression curves can be derived neglecting the marginal variability of wave steepness between 0.02 and 0.04, and spreadings between 15° and 20°. An overview of the regression curves for the different caisson geometries is shown, without the plot of the numerous scattered data, in fig.5 for the short-crested case : the nearly parallel curves show that different 3-D wave conditions induce a similar trend whatever geometry is

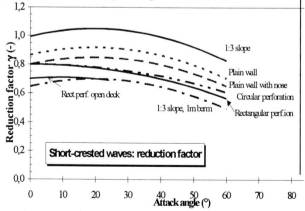

Fig.5 :Regression curves,3-D attacks to different caisson geometries.

considered. This result suggested a sort of "normalization" with regard to the geometry in order to define a unique γ function independent from the structural layout . The resulting fitting curves can be expressed in the form :

$$\gamma_\beta = 1.02 \cos^{1/3} (\beta) \qquad \text{for long-crested waves} \qquad (4)$$

$$\gamma_\beta = 0.92 \cos^{2/3} (10° - \beta) \qquad \text{for short-crested waves} \qquad (5)$$

as shown in fig: 6 together with the similar results obtained by de Waal and Van der Meer (1992) for slopes without crownwall under multi-directional waves. This study revealed that sloping and vertical coastal structures behave differently under 3-D sea-states. The obliquity effect over vertical structures, has a similar trend for long-crested and short-crested waves.

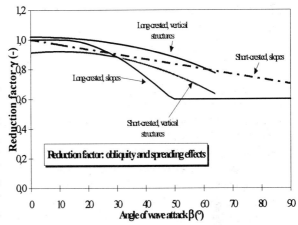

Fig 6: 3-D overtopping performance of sloping and vertical structures

In fact, even when waves attack the structure normally, wave short-crestedness gives no overtopping reduction on slopes, whereas a 10% reduction is observed for vertical-face caissons. The performance of the different caisson geometries has been estimated in fig. 7, where the γ_β and γ_σ factors are included within the x-axis, by means of correction factors ($\gamma_{geom.}$) to be applied to the response of the simple plain wall case . In table 1 a summary of all the $\gamma_{geom.}$ values is reported, together with the relative percentual differences in terms of γ and discharges, with reference to the plain wall case ($\gamma = 1.0$) . The slope directly connected to the caisson front wall produced a remarkable increase of overtopping, due to the large wave breaking at the wall toe . However, when a wide berm was added at the top of the slope very few waves managed to reach and overtop the caisson. The two perforated caissons achieved similar performances, even though

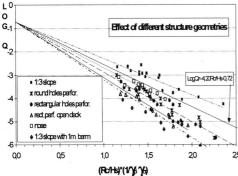

Fig 7 Influence of caisson geometries on wave overtopping.

Configuration	b geom	γ geom	‰(on γ)	‰ q(m/m)
Plain wall	-4,209	1,000	In relation to Plain wall case	
Wall with nose	-4,526	0,930	-7%	-27% / -46%
1:3 smooth slope	-3,613	1,165	17%	+84% / +220%
1:3 smooth slope with berm	-5,719	0,736	-26%	-77% / -94%
Round holes perforation	-4,794	0,878	-12%	-43% / -66%
Rectangular holes perforation	-4,960	0,850	-15%	-51% / -74%
Rectangular holes perforation with open deck	-5,502	0,765	-23%	-72% / -81%

Table 1 Influence of different structure geometries

the wall with rectangular holes wall has shown a slightly better overtopping response particularly under oblique wave attacks. The overall data set, can be then represented by the :

$$Q = 0.19 \exp(-4.20 R/(\gamma_\beta \, \gamma_\sigma \, \gamma_{geom}.))$$

The reliability of the formula can be expressed by the standard deviation $\sigma(4.20)=0.27$ assuming a normal distribution of the stochastic variable .

CONCLUSIONS

• The widely adopted exponential relation between the adimensional discharge and freeboard applies well also for 3-D conditions and a general equation can be defined accounting for all the calculated reduction factors, which can be useful for design purposes.

• Vertical coastal structures appear to have a slightly different response under 3-D and oblique waves if compared to sloping structures: slopes show a gentler and constant reduction, whereas caissons exhibit a reduction which starts at normal attack angles (0°). However the overtopping reduction under oblique long-crested wave attack is smaller for vertical structures (figure 6).

• The effect of multidirectionality can be given by a 10% freeboard reduction as compared to the traditional unidirectional case. Almost the same trend with the different wave conditions is shown for the various tested geometries

• The influence of structural configuration, as expressed in terms of freeboard reduction factors referred to the plain wall case, is summarized in table 1: a crest nose only gives a 7% reduction while a perforated front wall can improve the response to 12% (circular holes) or 15% (rectangular holes), or up to 23% with the opening of caisson top deck . A simple smooth submerged slope in front of the caisson instead increases the overtopping (+17% R_c increment), unless a large berm is placed at the slope crest (26% freeboard reduction) .

ACKNOWLEDGEMENTS

This 3-D model investigation has been partly financed by the UE under the (LIP) and the MAST II-MCS projects and partly by Dutch Rijkswaterstaat (TAW). The author gratefully wish to thank J.Van der Meer, J. Wouter (Delft Hydraulics), L. Franco and C. Restano (Politecnico di Milano) for their cooperation and advices .

REFERENCES

De Waal J.P. and van der Meer J.W.(1992), Wave runup and overtopping on coastal structures. Proc. XXIII ICCE, Venice, ASCE, N.Y., vol.2 .

Franco L., (1993), Overtopping of vertical breakwaters: results of model tests and admissible overtopping rates, MAST2-MCS project 1st project workshop, 13-14 Oct. 93, Madrid.

Franco L., De Gerloni, M. and Van der Meer, J. W., (1994), Wave overtopping at vertical and composite breakwaters, ASCE, proc. 24th ICCE, Kobe, Japan

Goda Y.(1985), Random seas and design of maritime structures .University of Tokyo Press.

A Monotonic Version of the Holly-Preissmann Scheme

V. GUINOT

L.H.F., Grenoble, France

INTRODUCTION

Consider first-order differential equation for advection:

$$\frac{\partial C}{\partial t} + u \frac{\partial C}{\partial x} = 0 \qquad (1)$$

where C = advected quantity (e.g. solute concentration); u = advection velocity; x, t = respectively space and time coordinates. When solving Eq. 1 numerically using finite differences or finite elements, at least two major difficulties arise, though Eq. 1 is apparently simple, namely :

(i) spurious diffusion and/or dispersion, due to a numerical scheme, the former leading to artificial damping of the computed solution, the latter resulting in oscillations ("wiggles") in the vicinity of moving fronts.

(ii) expression of boundary conditions in a way that is consistent with the numerical scheme used.

The present paper concentrates on improvements to the Holly-Preissmann (HP) scheme that allow better representation of steep fronts (elimination of overshots and wiggles) as well as consistent introduction of boundary conditions, which is equivalent to the ability to handle Courant numbers higher than 1.0.

It is first necessary to recall the principles of the HP scheme as shown in Fig. 1. The exact solution of eq. 1, $C(i \, \Delta x, (n+1) \, \Delta t) = C_i^{n+1}$, or the concentration sought at point B, is equal to the concentration at point A, $C(x_A, n \, \Delta t) = C_A^n$, which is the point of intersection of the abscissa $t = n\Delta t$ by the bakward characteristic drawn from B. The value of C_A^n can be found because all values C_i^n, (i = 1, ..., N) are known in time $n\Delta t$. In order to find C_A^n, the Holly-Preissman scheme uses a Hermitian third-order interpolating polynomial at time $n\Delta t$, between points i-1 and i:

$$C^n(\xi) = d_1 \xi^3 + d_2 \xi^2 + d_3 \xi + d_4 \qquad (2)$$

$$CX^n(\xi) = e_1 \xi^2 + d_2 \xi + d_3 \xi \qquad (3)$$

where $CX = \partial C/\partial x$, where ξ = auxiliary variable, $\xi = x -(i-1)\Delta x = (1 - \alpha)\Delta x$, where $\alpha = u\Delta t/\Delta x$ = Courant number.

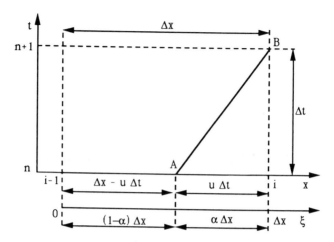

FIG. 1. Definition sketch of the characteristic-based, Holly-Preissmann Scheme

It is easy (see Holly and Preissmann, 1977) to derive the formulas :

$$C_i^{n+1} = a_1 C_{i-1}^n + a_2 C_i^n + a_3 CX_{i-1}^n + a_4 CX_i^n \tag{4}$$

$$CX_i^{n+1} = b_1 C_{i-1}^n + b_2 C_i^n + b_3 CX_{i-1}^n + b_4 CX_i^n \tag{5}$$

In essence, the Holly-Preissmann scheme solves two first-order partial differential equations, one for the sought variable $C(x,t)$ and one for its derivative $CX(x,t)$. This scheme is extremely convenient because it involves only two points at $n\Delta t$ level. Nevertheless, Hermitian interpolation is non-monotonic and this may be a disqualifying property for modellers, because of the appearence of wiggles and non-physical values (e.g. negative concentrations) in the computed profiles. On the other hand, it is not obvious, when using the Holly-Preissmann Scheme, to deal with boundary conditions, since space derivative CX is generally unknown at the boundary.
The subject of this paper is to propose solutions to these deficiencies, starting with boundary conditions.

BOUNDARY CONDITIONS
Consider Fig. 2, assuming that function $C(0,t)$ is known at the boundary. The Holly-Preissmann numerical scheme is an approximation to two first-order or to one second-order partial differential equations. Hence, for $\alpha < 1$, one needs two

(initial) conditions at point A where the backward characteristic originating at B intersects time level $n\Delta t$.

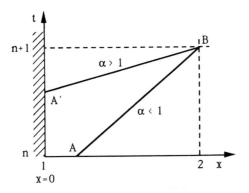

FIG. 2. Upstream boundary condition

If the above interpolation procedure is to be applied, one needs two boundary conditions: (C_1^n, CX_1^n) in order to interpolate C_A^n and CX_A^n values. These conditions should be independent while satisfying Eq. 1. Boundary condition $CX_1^n)$, which is usually unavailable from physical conditions, must be estimated. The most consistent approach is to use Eq. 1 and to rewrite it as

$$CX_1^n = \left.\frac{\partial C}{\partial x}\right]_{x=0}^{t=n\Delta t} = -\frac{1}{u_1^n}\left(\frac{\partial C}{\partial t}\right)_1^n \tag{6}$$

Second-order approximation in time for the $\partial c/\partial t$ derivative, such as

$$\left(\frac{\partial C}{\partial t}\right) \approx \frac{C_i^{n+1} - C_i^{n-1}}{2\,\Delta t} \tag{7}$$

leads to

$$CX_1^n = -\frac{C_1^{n+1} - C_1^{n-1}}{2\,u_1^n\,\Delta t} \tag{8}$$

On the other hand, using C_1^{n-1}, C_1^n and CX_1^{n-1} to approximate $(\partial C/\partial t)_1^n$ yields:

$$CX_1^n = \frac{2}{u_1^n\,\Delta t}\left(C_1^{n-1} - C_1^n\right) - \frac{u_1^{n-1}}{u_1^n}CX_1^{n-1} \tag{9}$$

Eq. 9, which is still second-order accurate, shows that if $C_1 = \text{const.}$, $u_1 = \text{const.}$, then, $CX_1^n = -CX_1^{n-1}$ and unavoidable oscillations appear. Using Eq. 7

does not yield such oscillations, but it cannot be used as an internal boundary condition (*e.g.* for looped network simulations), since one needs computational results at (not calculated yet) time level n+1 to compute the boundary condition for time level n. On this basis, first-order approximation with respect to time is preferred at the boundary:

$$CX_I^n = \frac{C_I^{n-1} - C_I^n}{u_I^n \Delta t} \tag{10}$$

Eq. 8 and Eq. 9 show that two different ways of estimating CX at the boundary, although of the same order of accuracy (namely second-order), are not always equivalent.

It should also be noted that the boundary problem is identical to the problem of Courant number $\alpha > 1$ (viz. Fig. 2, point A'). This approach, however, if applied to each computational point along the grid, leads to extensive artificial diffusion (see Holly and Rahuel, 1990).

INTRODUCTION OF MONOTONICITY

As Hermitian interpolation is non-monotonic, oscillations may appear in computed profiles. The purpose of this study was to eliminate such oscillations, by locally reshaping the interpolated profile so as to reach monotonicity.

To ensure monotonicity of the profile between points (i-1) and (i), the interpolated concentration derivative CX must be of the same sign over the interval [i-1, i]. This can be achieved by using (whenever necessary) modified values of CX_{i-1} and CX_i for the interpolation in Eqs. 2 and 3. The modified values will be denoted CXM_{i-1} and CXM_i hereafter.

First step : correction of the signs of space derivatives
Obviously, if the signs of CX_{i-1} and $(C_i - C_{i-1})$ are opposite, monotonicity cannot be achieved. CXM_{i-1} is thus set to zero if $\left[CX_{i-1} (C_i - C_{i-1}) \right] < 0$; otherwise, $CXM_{i-1} = CX_{i-1}$.
The same treatment should be applied to CX_i.

Second step : correction of the values of space derivatives
Having $CXM_{i-1} (C_i - C_{i-1}) > 0$ may not be a sufficient condition to ensure monotonicity. The space derivative CX is estimated by a second-order polynomial, hence its sign may change over [i-1, i], thus destroying monotonicity. If the case, the non-zero value of CXM is modified such as the CX interpolation parabola keep the same sign within [i-1, i].

TEST APPLICATION

The modified scheme, using monotonicity modifications, was applied to the advection of a step function profile by a constant velocity field (u = 1 m/s). The cell width is regular, Δx = 1 m. The calculation timestep was constant (Δt = 0.8 s), so that α = 0.8. The simulation was performed over a 16 s period (20 timesteps). Figs. 3 a-c show the results given by linear interpolation, for C_A^n original (HP) Hermitian interpolation and Hermitian interpolation with monotonicity constraints, while boundary condition dicretization is defined respectively by Eq. 8 (second-order discretization with respect to time using three levels in time), Eq. 9 (second-order discretization with respect to time, using two time levels) and Eq. 10 (first-order time discretization, using two time levels).Wiggles induced by the boundary condition are eliminated by the modified, monotonic scheme, while they are propagated over the entire computational grid by the original, non-monotonic Holly-Preissmann scheme. Comparing Figs. 3-b and 3-c shows that boundary conditions may lead to very different computational results, depending on the way they are introduced - even when treated with the same accuracy.

CONCLUSION

The problems of boundary condition introduction and monotonicity in the Holly-Preissmann scheme have been presented. Various ways of introducing boundary conditions are also presented and computational examples show that the boundary condition problem should be treated with care. Monotonicity can be achieved, by locally reshaping the Hermitian interpolation used by the scheme. Artificial oscillations in computed profiles are eliminated, without noticeably degrading the performances of the scheme.

ACKNOWLEDGEMENTS

This work was carried out at LHF (Laboratoire d'Hydraulique de France). It was inspired by the TSAR project, involving a cooperation between LHF, STC (Service Technique Central des Ports Maritimes et Voies Navigables, Compiègne, France) and LNH (Laboratoire d'Hydraulique de France - EDF Chatou).
The author would like to thank Dr J. A. Cunge from LHF for comments and suggestions.

REFERENCES

Holly, F.M. Jr., and Preissmann, A. (1977). "Accurate calculation of transport in two dimensions." *J. Hydr. Div.*, ASCE, 103 (11), 1259-1277.
Holly, F.M. Jr., and Rahuel, J.-L. (1990). "New numerical/physical framework for mobile-bed modelling. Part 1: Numerical and physical principles." *J. Hydr. Res.*, IAHR, 28 (4), 401-416.

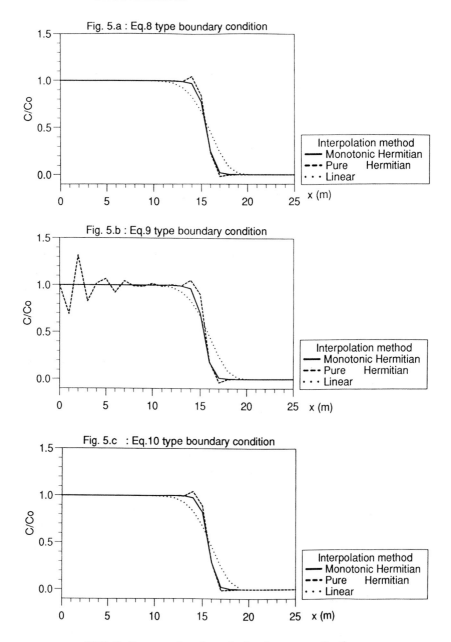

FIG. 3. Computational results for the test application

TURBULENCE MODELLING OF OPEN-CHANNEL FLOW IN PIPES OF CIRCULAR CROSS-SECTION WITH A FLAT BED

C. Hoohlo
Research Student
Department of Civil Engineering,
University of Newcastle upon Tyne, UK.

SUMMARY

The numerical model is based on the Semi-Implicit Method for Pressure-Linked Equations (SIMPLE), and computes the flow on a Cartesian grid. It uses a non-linear k-ε turbulence model to compute the three-dimensional flow using wall functions. The model boundary conditions are modified to reflect the effects of the corners, the curved side-wall, and a roughened bed on the flow.

The computed velocity, and turbulence parameter distributions are close to empirical ones and those obtained by experiment. The model also accurately predicts the secondary flow.

1   INTRODUCTION

To design against localised degradation of the open-channel boundary, or sediment deposition in low flows, the local flow conditions are required. Refined flow modelling is aimed at reproducing these channel-wide variations of the velocity, and boundary shear stresses by solving the governing Navier-Stokes equations.

Figure 1.1 illustrates the geometry of the pipe of circular cross-section with a flat bed. The flow studied is longitudinally uniform and three-dimensional - with secondary flow in the yz-plane - due to the flat bed, side-walls and, free-surface (Hoohlo 1994).

1.1   Governing Equations

In tensor notation, the time-averaged Navier-Stokes equations (Eqn 1) for steady flow

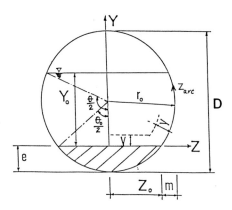

Figure 1.1 The Channel Cross-Section

$$U_i \frac{\partial U_i}{\partial x_j} = -\frac{1}{\rho}\frac{\partial P}{\partial x_i} + v\frac{\partial_{\cdot}^2 U_i}{\partial x_j \partial x_j} - \frac{\partial \overline{u_i' u_j'}}{\partial x_j} + F_i \qquad (1)$$

I   II   III   IV   V

contain the unknown Reynolds stress terms $(-\rho\overline{u_i' u_j'})$. ρ = density, and v = the kinematic viscosity. The index i = 1, 2, 3 represents X- (longitudinal), Y- (vertical), and Z- (transverse) directions respectively. U_i = the time-mean velocity in the ith direction, and P = the time-mean pressure. F_i is the body force per unit mass acting in the ith direction (e.g. $F_1 = \rho g S_o$, where g = gravitational acceleration, and S_o = bed slope).

The total stress term (Terms III and IV) is represented by (see e.g. Nezu and Nakagawa 1993)

$$\tau_{ij} = -\rho\overline{u_i' u_j'} = 2\rho v_t \overline{S_{ij}} - \frac{2}{3}\rho k \delta_{ij} \qquad (2)$$

where v_t = the eddy viscosity (noting that $v_t \gg v$ in turbulent flow), $\delta_{ij} = 1$ when i = j, and = 0 otherwise, k is the turbulent kinetic energy, and $\overline{S_{ij}} = 1/2[\partial U_i /\partial x_j + \partial U_j /\partial x_i]$ is the mean rate of strain. Boussinesq's eddy viscosity hypothesis represents the Reynolds stress as a linear function of the mean rate of strain.

Using the principles of Material Frame Indifference (MFI) and Determinism, Speziale (1987) derived a non-linear expression for the stress, τ_{ij}, quadratic in $\overline{S_{ij}}$. A similar but more numerically stable expression had been developed by Baker and Orzechowski (1983)

$$\tau_{ij} = -\frac{1}{3}\overline{u'_k u'_k}\, a_i\, \delta_{ij} + c_4\frac{k^2}{\varepsilon}\overline{S_{ij}} + c_2\, c_4\frac{k^3}{\varepsilon^2}\overline{S_{ik}}\,\overline{S_{kj}} + \dots \tag{3}$$

where a_i are coefficients admitting anisotropy ($a_1 = 0.94$, $a_2 = a_3 = 0.56$). $c_2 = 0.067$, and $c_4 = 0.068$.

1.2   Turbulence Modelling

In the k-ε model, the eddy viscosity is given as

$$v_t = c_\mu \frac{k^2}{\varepsilon}, \tag{4}$$

where c_μ, the empirical constant, ($\equiv c_4$) $= 0.09$ is used. The turbulent kinetic energy, k, ($=\frac{1}{2}(\overline{u'_1 u'_1} + \overline{u'_2 u'_2} + \overline{u'_3 u'_3})$)), and the rate of energy dissipation, ε, are modelled from their respective simplified transport equations

$$U_i\frac{\partial k}{\partial x_i} = \frac{\partial}{\partial x_i}(\frac{v_t}{\sigma_k}\frac{\partial k}{\partial x_i}) + v_t\,(\frac{\partial U_i}{\partial x_j} + \frac{\partial U_j}{\partial x_i})\frac{\partial U_i}{\partial x_j} - \varepsilon \tag{5}$$

$$U_i\frac{\partial \varepsilon}{\partial x_i} = \frac{\partial}{\partial x_i}(\frac{v_t}{\sigma_\varepsilon}\frac{\partial \varepsilon}{\partial x_i}) + c_{1\varepsilon}\frac{\varepsilon}{k}P_K - c_{2\varepsilon}\frac{\varepsilon^2}{k}. \tag{6}$$

The Prandtl numbers σ_k and σ_ε are 1.0 and 1.30 respectively, and the empirical constants are $c_{1\varepsilon} = 1.44$, $c_{2\varepsilon} = 1.92$. $P_k =$ production of k. Eqns 3 - 6 represent the full non-linear k-ε model. This model is able to model the normal stress anisotropy responsible for turbulence-driven secondary flow.

2   NUMERICAL MODEL

The numerical computation is based on the SIMPLE technique (Patankar and Spalding 1972, Van Doormal and Raithby 1984) carried-out on a Cartesian grid. Wall functions

(Launder and Spalding 1974), are used to specify the rigid-wall and free-surface boundary conditions (Hoohlo and Nalluri 1993, and Hoohlo 1994).

Using the general form

$$U_j \frac{\partial(\rho \phi)}{\partial x_j} = \frac{\partial}{\partial x_j}(\Gamma_\phi \frac{\partial \phi}{\partial x_j}) + S_\phi$$

[Convection] [Diffusion] [Source]   (7)

for Eqns 1, 5, and 6, where $\phi \equiv U_i$, k, or ε and $\Gamma_\phi = \nu_\phi/\sigma_\phi$ = diffusivity, the additional diffusion terms that result from using Eqn 3 are 'lumped' into the source term, S_ϕ.

3   MODEL RESULTS AND DISCUSSION

Figures 3.1 and 3.2 show results for the mild bed slope $S_o = 9.27 \times 10^{-4}$, bed thickness (e),

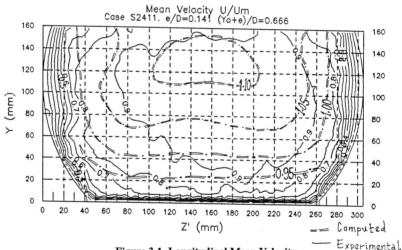

Figure 3.1 Longitudinal Mean Velocity   == Computed — Experimental

$e/D = 0.141$, and flow depth ratio $(Y_o+e)/D = 0.666$, each variable compared to their respective empirical two-dimensional profiles given in Nezu and Nakagawa (1993). Although the numerical primary velocity plot is smoother than the experimental one, the patterns are similar.

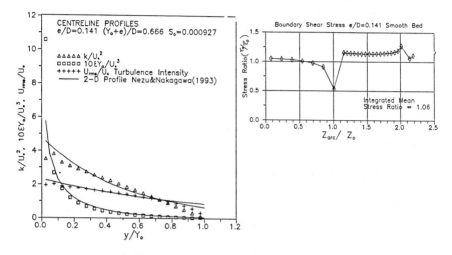

Figure 3.2 Normalised Turbulence Parameters and Boundary Shear Stress

4 CONCLUSIONS

Model predictions of the primary mean velocities, the secondary flow and, the centreline turbulence parameters (k, ε, and $U_{rms} = \sqrt{u_i' u_i'}$) are close to the experimental and empirical distributions. As in rectangular channels, the predicted local boundary shear stress decreases from the centreline along the bed and minimises at the corner. Hoohlo (1994) lists the limitations of the model (e.g. for the mesh size when $(Y_o + e)/D < 0.5$) which affect the secondary flow and boundary shear stress predictions by the model.

5 ACKNOWLEDGEMENTS

The author gratefully acknowledges the guidance of his supervisor, Dr C. Nalluri, and thanks the Lesotho Highlands Developement Authority for the financial support that enabled this study.

6   REFERENCES

Baker, A.J. and Orzechowski, J.A. (1983). "An Interaction Algorithm for Three-Dimensional Turbulent Subsonic Aerodynamic Juncture Region Flow", **AIAA Journal**, 21(4), 524-533.

Hoohlo, C., and Nalluri, C. (1993). "Velocity, Turbulence and Boundary Shear Distributions in Pipes of Circular Cross-Section with a Flat Bed: Turbulent Flow", **Numerical Methods in Laminar and Turbulent Flow: Proceedings of the 8th International Conference**, (Ed. C. Taylor), Swansea: Pineridge Press, Pt. 1(2), 230 - 240.

Hoohlo, C. (1994). "A Numerical and Experimental Study of Open-Channel Flow in a Pipe of Circular Cross-Section With a Flat Bed", **Unpublished PhD Thesis**, University of Newcastle upon Tyne.

Launder, B.E. and Spalding, D.B. (1974). "The Numerical Computation of Turbulent Flows", **Computer Methods in Applied Mechanics and Engineering**, 3, 269-289.

Nezu, I and Nakagawa, H. (1993). **Turbulence in Open-Channel Flows**. Delft: A.A. Balkema.

Patankar, S.V. and Spalding, D.B. (1972). "A Calculation Procedure for Heat, Mass and Momentum Transfer in Three-dimensional Parabolic Flows", **International Journal of Heat and Mass Transfer**, 15, 1787-1806.

Speziale, C.G. (1987). "On Non-Linear k-l and k-ε models of Turbulence", **Journal of Fluid Mechanics**, 178, 459-475.

Van Doormal, J.P. and Raithby, G.D. (1984) "Enhancements of the SIMPLE Method for Predicting Incompressible Fluid Flows", **Numerical Heat Transfer**, 7, 147-163.

A MODEL TO INVESTIGATE THE ROLES OF TEXTURE AND SORTING IN BED ARMOURING.

BARRY J.E. JEFCOATE
University of Aberdeen, Scotland.

INTRODUCTION

In conjunction with careful physical experiments the computer has proved itself to be a useful research tool in studying the micro-scale processes of sediment transport. Unfortunately, the limits of computer power and available time impose restrictions upon the level of detail which can be included in a model. These restraints necessitate a certain degree of compromise and abstraction. Existing models (e.g. Wiberg & Smith [1985], Sekine & Kikkawa [1992], Jiang & Haff [1993]) achieve this by one or more of the following methods: constraining the geometry to two dimensions; using single grain sizes or discrete distributions; rebounding grains over artificial, unrepresentative bed surfaces, or only considering the motion of single grains. The purpose of this paper is to introduce an improved model to investigate the relationship between the three dimensional arrangement of multi-sized material in the bed and the transported bedload.

A DESCRIPTION OF THE MODEL

The model is a discrete element model, in that the behaviour of the bed as a whole is modelled by considering the interaction of individual particles. The conceptual structure of the model is that of the entrainment, transportation, deposition cycle.

Bed Preparation.

In order to represent the three dimensional phenomena of sorting and texture successfully the model is also three dimensional. Grains are approximated by spheres for two reasons. Firstly, and most importantly, to keep the geometrical calculations within reasonable bounds. Secondly, the behaviour of a sphere is the same in any orientation to a flow (Carling et.al. [1992]). From now on the term 'grain' will be used to refer to one of these spherical particles. An initial bed (figure 1) is generated by taking grains from a predetermined, continuous size distribution and dropping them one at a time at random x & y co-ordinates onto a foundation of intersecting spheres. The grain is allowed to rebound from the spherical surfaces of the foundation layer, or other grains, until its energy is damped out and it forms a stable, geometrically consistent unit with the last three supports it encountered. An exact position for the grain is calculated at the apex of a tetrahedron with side lengths equal to the sum of the radii of the contacting spheres. Friction and elastic energy losses are modelled by a rebound coefficient. The unworked bed is laid down under dry conditions, as it would be in a physical flume experiment, so losses due to viscosity and lubrication at impact are assumed to be negligible.

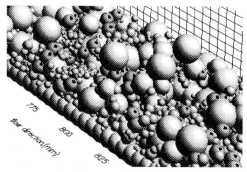

Figure 1. A small section of a simulated, unworked, bed showing a continuous distribution of grain sizes and obvious texture.

Entrainment.

Each grain in the bed surface is subjected to a series of tests which determine whether it is to be eroded under the given flow conditions and bed state. The hierarchy of the tests is;

    i). Only grains with a negative downstream friction angle are considered.

    ii). A grain will be considered for entrainment as long as fewer than a prescribed number of grains are dependent upon it for their stability.

    iii). A grain exposed to a flow experiences the forces of weight, buoyancy, drag, and lift. Those grains which a moment analysis shows to be unstable are entrained. As are any grains which have had a support removed and cannot form a stable structure with any of their immediate neighbours.

Transportation.

After it is entrained into the flow the grain follows a trajectory which is calculated by considering the forces acting upon the grain. The equation of motion of a grain is shown in (1).

$$\rho(\sigma_s + C_m)V_p \frac{d\mathbf{u}_p}{dt} = \rho(\sigma_s - 1)V_p\mathbf{g} - 0.5\rho C_d A_p|\mathbf{u}_r|\mathbf{u}_r \qquad (1)$$

The left hand side is an inertial term; ρ is the density of water, σ_s is the ratio of the particle density to that of water and V_p is the volume of the particle. The product of these three quantities is the mass of the particle. C_m is an added mass coefficient (0.5 for spheres) to take into account the force required to accelerate the fluid as the grain accelerates through it. The grain's acceleration is given as the rate of change of its velocity, $\mathbf{u}_p$ with respect to time, t. In order, the terms on the right are, a combined weight and buoyancy term (a reduced mass multiplied by the acceleration due to gravity, $\mathbf{g}$) and a drag force. In common with Sekine & Kikkawa [1992] the drag force is calculated from a drag coefficient, C_d, which depends on the particle Reynolds number as suggested by Rouse [1938]. The other important variables in the drag force calculation are the projected area of the spherical grain, A_p, and the velocity of the grain relative to the flow, $\mathbf{u}_r$.

In addition, there is also a lift force, $\mathbf{F}_l$. Lift is calculated from an expression derived by Wiberg & Smith [1985] with an average value of the lift coefficient, C_l. Given the velocity of the grain relative to the fluid at its top, $\mathbf{u}_t$, and at its bottom, $\mathbf{u}_b$,the lift force is calculated as follows;

$$\mathbf{F}_l = 0.5\rho \cdot A_p \cdot C_l \left[\mathbf{u}_t^2 - \mathbf{u}_b^2\right] \qquad ,C_l \approx 0.2 \qquad (2)$$

The velocity field currently being used is a simple, time averaged, one dimensional flow profile. At all points above the bed the differential flow velocity across the grain is thus zero, resulting in a zero lift force once the grain has been entrained. Lift forces throughout the trajectory will be introduced at a later date with the use of a more sophisticated flow model. However, numerical experiments by Sekine & Kikkawa [1992] have shown that trajectories of grains remain relatively unaffected if lift forces are neglected in conjunction with those forces which arise from local turbulence.

Deposition.
During each successive collision the grain imparts some of its momentum to the bed. As a result of this the saltating grain loses kinetic energy. A grain will continue to rebound until it no longer possesses sufficient energy and is trapped in a hollow in the bed. The final resting position will be on three principal supports. If the grain is deemed to be stable in this position by a moment analysis it will remain until there is a change in the local flow conditions or the arrangement of the neighbouring grains is disturbed.

RESULTS AND DISCUSSION.
The model has been used to simulate bedload transport over a bed of a continuous grain size distribution, 1 m. in length (figure 1).

Sediment Transport Rate.
All the transported material was entrained from the bed and rebound angles were calculated by applying 3 dimensional geometry at each point of impact with the bed. Whether a grain was subsequently re-entrained after being deposited on the bed depended entirely upon the stability and exposure of its new position in relation to its immediate neighbours. There was no feed of new material, so the process was one of static armouring.

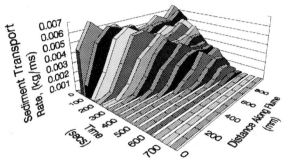

Figure2.
Transport rates calculated for a mean flow vel. U=0.3m/s, and corresponding shear velocity, U*=0.04m/s.

The sediment transport rate was calculated at 10 equally spaced points along the model domain at 50 second intervals. The resulting spatial profiles of sediment transport rate are plotted in figure 2. At the outset, the sediment transport rate rises rapidly to between 0.005kg/ms and 0.007kg/ms which is fairly uniform along the whole length of the flume. The reason for the initial, sharp increase in sediment transport rate can be attributed to the initially unworked state of the bed. Some grains occupy precarious positions from which they are easily entrained.

A reduction in the availability of sediment for transport was observed over time, and also with distance along the flume. This is evident as a reduction in the peak value of the sediment transport rate as it advanced downstream. This is associated with a region of bed armouring developing at the upstream end of the flume and progressing downstream. In comparison with results of equivalent static armouring experiments carried out in the laboratory (Tait et. al. [1992]), the calculated sediment transport rate declines more rapidly. Although it takes 750 seconds for the sediment transport rate calculated at 1 metre to be recorded as nearly zero, at 100 mm from the head of the model this delay is only 250 seconds. In the experiments of Tait et. al.[1992] the sediment transport rate was measured using a trap located at 8 metres from the upstream end of the flume. This difference in scale could account for up to an order of magnitude on the time scale. The fact that the calculated sediment transport rate has fallen to zero is due to the bed reorganising itself into a more stable configuration. This is associated with an increase in the critical shear stress for the bed.

Turbulent Uncertainty in the Flow Model.
The flow model used in these preliminary experiments is one dimensional, steady and uniform. The drag force on a grain acts only in the stream wise direction and does not vary. It is anticipated that the effect of introducing a probability distribution to model the effect of turbulence (after Grass [1970]) will be to reduce the peak calculated sediment transport rate and to increase the time over which transport occurs. It is recognised that turbulence has an important role in entrainment and bed development. However, to date, while investigating the part that the organisation of the bed plays in sediment availability it has been expedient to keep the flow model as simple as possible.

Grain Size Distribution.
The grain size distributions of the original unworked bed and the material which passed the end of the model are shown in figure 3. All the grain sizes of the original bed mixture are present in the trap, though in different proportions. Early in the simulation the size distribution of the material caught in the trap bears a similarity with the material in the parent bed.

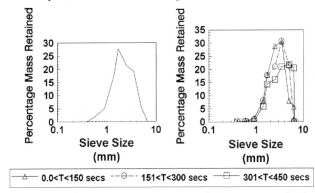

Figure3. The grain size distribution of the original bed mixture (left) and the material which was trapped at 1 metre over 3 time intervals (right).

The tendency is for the modal value of the transported mixture to increase with time, and for the shape of the distribution to become less peaked, developing a bias towards those grains which are finer than the mode. This can be seen as a reduction in the gradient of the fine leg of the distribution and the development of a 'shoulder'. This departure of the grain size distribution from the original mixture reflects changes in the composition of the bed surface.

At the outset of the experiment it is the large grains which are more prominent in the bed, and so early observations of transported material are biased towards the coarse end of the distribution. Removal of the coarse surface grains exposes the finer grains that had benefited from the shelter and protection of their larger neighbours, resulting in the skewing of the grain size distribution towards the finer fractions. This change in the nature of the transported material suggests that it is unsatisfactory to assume that all grains of the same size will have the same threshold of motion. The relative exposure of the grain, as well as the grains size, is an important determinant of its susceptibility for entrainment. Considering each grain individually overcomes the difficulties in determining a critical flow condition for the whole bed.

CONCLUSIONS

This model has many inherent advantages, it can be used to investigate the three dimensional effects on the transport of a continuously distributed non-cohesive sediment mixture under different flow conditions. Because of its physical basis the model can also be used to look at the texture of the surface, as well as the arrangement of grains throughout the bed. Although still in its infancy, the encouraging results in this paper show that the model predicts a reduction in the sediment transport rate with both time and distance, and also a change in the composition of the transported material. The intention now is to use the model to investigate the influence of different mixtures on these phenomena, in order to improve the understanding of the relationship between grain size distribution, the arrangement of the bed, and the bed load transport rate.

REFERENCES

Carling, P.A., Kelsey, A., & Glaister, M.S., Effect of bed roughness, particle shape and orientation on initial motion criteria, in *Dynamics of gravel bed rivers*, ed. Billi, P., Hey, R.D., Thorne, C.R., & Tacconi, P., John Wiley & Sons Ltd., 1992.

Grass, A.J., Initial instability of fine bed sand, *Journ. Hydr. Div. Proc. A.S.C.E.*, 96(HY3), pp.619-632, March 1970.

Jiang, Z., & Haff, P.K., Multiparticle simulation methods applied to the micro mechanics of bed load transport, *Water Resour. Res.*, 29(2), pp. 399-412, February 1993.

Rouse, H., *Fluid mechanics for hydraulic engineers*, Dover, New York, 1938.

Sekine, M., & Kikkawa, H., Mechanics of Saltating Grains II, *Journ. Hydr. Eng*, 118(4), April 1992.

Tait, S.J., Willetts, B.B., & Maizels, J.K., Laboratory observations of bed armouring and changes in bedload composition, in *Dynamics of gravel bed rivers*, ibid.

Wiberg, P.L., & Smith, J.D., A theoretical model for saltating grains in water, *Journ. Geophys. Res.*, 90(C4), pp.7341-7354, July 1985.

NUMERICAL CALCULATION OF PRESSURE FLUCTUATION IN RECIRCULATION FLOWS USING DISCRETE VORTEX MODEL

By M. R. KAVIANPOUR[1]
Department of Civil and Structural Engineering,
UMIST, P.O. Box 88, Manchester

ABSTRACT

The purpose of this study is to apply the discrete vortex model to determine pressure fields in the reattaching zone of the flow past a normal plate situated in a duct. The plate is of negligible thickness which is aligned normal to a uniform approaching stream. The plate spanned the entire width of the duct.

In this model, the motion of the free shear layer is represented by elemental vortices. The complex potential is derived through a Schwartz-Christoffel transformation of the physical plane. In addition to considering the combination and the cancellation of vortices, the circulation of them was also allowed to decay with time in order to take partial account of the real fluid behaviour. A large number of computations have been performed until the fluctuation of velocities appeared to be statistically stationary. These calculations yield predictions of the time-mean and r.m.s. values of the velocity and pressure fluctuations in the separation zone which, have been compared with the experimental results.

INTRODUCTION

Some of the flows in which reattachment occurs are those over a normal wall,

---

1. Academic staff of Department of Civil Engineering,
   Amirkabir University of Technology, Teheran, Iran.

through a partially closed gate valve, past upstream and downstream-facing steps, and in sudden changes in conduit sections. Major problems encountered in such applications are cavitation, structural vibration and fatigue failure.

Recent investigations show that pressure fluctuations could be responsible for cavitation to occur sometimes even when the mean pressure is well above vapour pressure [3]. The fluctuating pressures may cause vibrations, fatigue of materials and large instantaneous pressures close to the vapour pressure of liquid. Damage due to pressure fluctuations has been serious at a number of structures, such as, spillways, stilling basin and reattaching flows [6,7]. Therefore, it should be an important part in the design of hydraulic structures [5]. The present result of pressure fluctuation is entirely experimental; here an attemp is made to predict the intensity of pressure fluctuation.

In this work discrete vortex model have been used to predict the pressure fluctuations arising from a normal plate placed in a duct. This model is popular in view of its relative simplicity and of its ability to produce satisfactorily some of the features of real flows. One of the most important application of the discrete vortex model is in analysing the flow past bluff bodies. In this approach reviewed by Clements and Maull [1] and more recently by Sarpkaya [8] and Lewis [2], the shear layer shed from the separation point of the bluff bodies is approximated by arrays of line vortices. Inviscid elemental vortices representing the shear layer are introduced at the separation point at every time interval Δt. The convection velocity of each vortex which is the sum of the velocity of the irrotational potential flow around the body and that induced at vortex centre by all other vortices is computed. Then the vortices are convected according to their computed velocities to the new positions using a suitable scheme. Again a new vortex is introduced and thus, the movement of these discrete vortices provides a simulation of the wake pattern downstream of the body. In the convection of vortices their individual identities are maintained except in circumstances in which they approach each other or a solid boundary too closely.

MATHEMATICAL DESCRIPTION OF MODEL

The model proposed here is an infinite two-dimensional channel of depth D containing a perpendicular plate of height h. The complex potential of the flow is found by using the Schwarrans-Christoffel transformation which maps the interior of the duct of the z-plane into the upper half of the t-plane by introducing a source-sink combination corresponding the points of infinity. The conformal transformation is given by;

$$z = \frac{2D}{\pi} \tanh^{-1} \left(\frac{t^2-1}{\beta^2-1} \right)^{\frac{1}{2}}$$ (1)

where β which represents the position of source and sink and corresponds to the transformed points at infinity.

To simulate the velocity shed from the edge of the plate, discrete vortices are introduced into the flow at this point. Since these elemental vortices are continually converted away by the mainstream flow and their own induced velocity, new discrete vortices are generated at discrete time Δt to continue the simulation process. The subsequent path formed by their co-ordinates represents the shear layer shed from the edge. Thus the complex potential in the transformed plane is given by the sum of complex potential due to placing of source and sink $W_u(t)$ plus those induced by vortices $W_v(t)$.

$$W(t) = \frac{DU}{\pi} [Log(t+\beta) - Log(t-\beta)] + \frac{i}{2\pi} \{ \sum_{j=1}^{n} K_j [Log(t-t_j) - Log(t-t_{j0})] \}$$ 2

which includes the potential due to an image system introduced to prevent flow across the boundary. In this relation t_j represents the vortex position in the transformed plane corresponds to that at z_k, k_j is the strength of the vortex and U is the magnitude of the flow velocity through the channel.

COMPUTATION OF VELOCITY AND PRESSURE FIELDS
Time-mean and r.m.s. value of the fluctuating parts of the velocity computed from a time period when the number of vortices appear to be statistically stationary. The pressure distribution at any section downstream of the plate is obtained by applying Bernoulli equation and was determined in the form of a pressure coefficient;

$$Cp = \frac{P-P_0}{\frac{1}{2} \varrho U^2} = 1 - \frac{2}{U^2} \frac{\partial W}{\partial t} - \frac{1}{2} \frac{1}{U^2} \left| \frac{dW}{dz} \right|^2$$ (3)

in which P_0 is the pressure in the undisturbed stream and $\partial w/\partial t$ is the unsteady term evaluated from potential equation. The distribution of r.m.s. value of pressure fluctuations is also presented in the form of $P'^2/(0.5\rho U_c^2)$ and determined by;

$$Cp_{mean} = \frac{\Sigma Cp . \Delta T}{\Sigma \Delta T}$$

$$Cp' = \frac{U^2}{U_c^2} [\frac{\Sigma Cp^2 . \Delta T}{\Sigma \Delta T} - (Cp_{mean})^2]^{\frac{1}{2}} \tag{4}$$

where U_c is the velocity due to the vena contracta. The computation of the mean and r.m.s. value of the velocity and pressure started from when the number of vortices inside the flow became stable.

RESULTS AND COMPARISONS

The experiments were carried out to determine the pressure field downstream of the plate in the hydraulic laboratory of the department of civil engineering at Umist. The time-mean and r.m.s. values of pressure fluctuations were computed and the results were compared with those obtained by experiments. Fig.1 shows the experimental set up. The distribution of vortices representing the reattaching region is represented in Fig.2. The r.m.s. value of pressure fluctuation beneath the recirculation region were presented in Fig.3. It shows a reasonable agreement specially around the recirculation area, but there is a difference with the experimental results downstream of the reattaching point. These disagreement can be related to the three dimensional nature of the flow, while the model is a two dimensional. The maximum pressure fluctuation in the shear layer represented in Fig.4 and compared with those of experiments. Fig.(6) shows the time-mean velocity distribution downstream of the plate. According to the graph, the length of the standing eddy has been computed around Xr/h=10 for H/D=0.2 which is near to the results of Narayanan [4] and the present experiments which is around 12H for H/D=0.1 and around 8-9H for H/D=0.2. Mean pressure distribution is presented in Fig.(5) which also shows a close agreement with the experimental results.

CONCLUDING REMARKS

The discrete vortex model has been utilized with some success to simulate the separation zone over a two dimensional flat plate with finite thickness. A simple and effective procedure as recommended by Kiya et al was employed to represent the effect of viscosity. The present calculation yield reasonable prediction of the time-mean and r.m.s. values of the velocity and pressure fluctuations except in the region downstream of the reattaching point which could be related to the nature of the flow which is three dimension.

The author would like to thank Dr. R. Narayanan for his general supervision of this work, which was carried out in the Civil Engineering Department of UMIST.

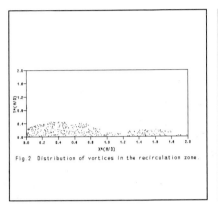

Fig.2 Distribution of vortices in the recirculation zone.

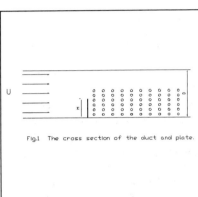

Fig.1 The cross section of the duct and plate.

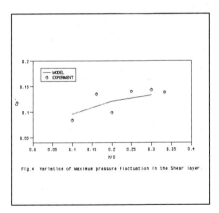

Fig.4 Variation of Maximum pressure Fluctuation in the Shear layer.

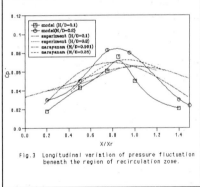

Fig.3 Longitudinal variation of pressure fluctuation beneath the region of recirculation zone.

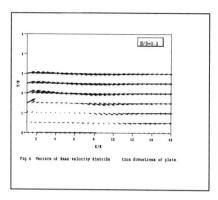

Fig.6 Vectors of mean velocity distribu   tion downstream of plate

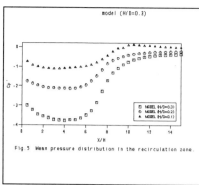

Fig.5 Mean pressure distribution in the recirculation zone.

REFERENCES

1. **Clements R. R. and D. J. Maull, 1975**
"The representation ofsheets of vorticity by discrete vortices", Prog. Aero. Sci., Vol. 16, p. 129-146

2. **Lewis R. I., 1987**
"Recent developments and Engineering application of the vortex cloud method", Computer Methods in Applied Mechanics and Engg., Vol. 64, p.153-176.

3. **Lopardo R., I. De lio, G. Vernet, 1987**
"The Role of Pressure fluctuations in the design of hydraulic structures", edited by R. Kiya and M. Albetson, Colorado state University, Fort Collins, p. 161-175.

4. **Narayanan R., 1970**
"The Reattaching flow downstream of a leaf gate", A thesis submitted to the Dept. of Mech. Engg., Brunel Univ., in partial fulfilment of the degree of ph.D.

5. **Narayanan R., 1984**
"The Role of Pressure fluctuation in Hydraulic Modelling", symp. on scale effect in modelling hydraulic structures, Esslingen, Germany, p.1.12-1 to 1.12-6.

6. **Narayanan R. and J. Reynolds, 1968**
"Pressure Fluctuation in Reattaching flow", ASCE, Hy., Vol. 94(6), p.1383-1398.

7. **Narayanan R. and J. Reynolds, 1972**
"Reattaching flow downstream of leaf gate", ASCE, Hy., Vol. 98(5), p.913-934.

8. **Sarpkaya T., 1989**
"Computational methods with vortex-The 1988 freeman scalar lecture", J. Fluid Engg., Vol. 111, p.5-55.

Sediment Budget for a Lock-Regulated Section of Neckar River at Storm Events

ULRICH KERN
Institute of Hydraulic Engineering, University of Stuttgart, Germany

ABSTRACT

At two storm events, the sediment budget was calculated for a lock-regulated section of Neckar River near Lauffen, Germany from concentration of suspended particulate matter (SPM) measured at its upstream and downstream boundaries. Sampling points proved to be representative, as their SPM concentration showed deviation from flow-weighted cross-sectional average by about 1%. Difference between outflowing and inflowing mass of SPM was about 32,000 tons for the first flood and about 24,000 tons for the second storm event. This is due to considerable net erosion of channel sediments from the regarded river-lock, especially in its lower reach.

INTRODUCTION

Lock-regulation of rivers decreases flow velocity and therefore, enhances settling of fine-grained suspended solids in the lock-chambers at low discharge. Due to sewage or industrial effluents in the catchment, the trapped sediments may be heavily loaded by trace pollutants (for Neckar River, see Förstner and Wittmann, 1981; Kern and Westrich, 1995a). During storm events, resuspension of sediments from backwater sections may lead to remobilization of sediment-stored contaminants. The purpose of this study is to quantify the net resuspension of sediments from a river-lock section at two flood events. Accordingly, mass balances of SPM are calculated from field data.

FIELD SITE

The regarded 10.9km long river-lock section of Neckar River is enclosed by hydraulic structures, upstream near Besigheim and downstream near Lauffen (Fig. 1). At Lauffen, the discharge of Neckar River ranges from 14.1m³/s to 1650m³/s at an average (MQ) of 88.5m³/s. Annual sediment yield from the catchment area (A_D) of 7916km² varies between 7.4tons/km² and 70.9tons/km². At its mouth near Besigheim, A_D and MQ of Enz River is 2230m² and 20.8m³/s, respectively.

DATA SET

During flood events in Dec. 1993 and in Apr. 1994, water samples were collected from Neckar River at stations BN, KN and LN and from Enz River at BE (Fig. 1). Stations LN and BN are located at tailwater position of the respective hydraulic structure. The samples were taken from the centroid of the flow near the water surface using buckets and stored in 1liter plastic bottles. Suspended particulate matter was concentrated and gravimetrically measured by filtration of the samples using 0.45-μm membrane filters (Sartorius, Göttingen).

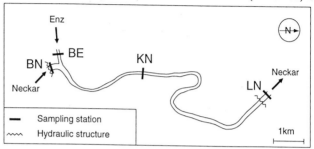

Fig. 1:   The lock-regulated section of Neckar River showing location of the sampling stations: Enz River at Besigheim (BE) and Neckar River at Besigheim (BN), Kirchheim (KN) and Lauffen (LN).

Discharges at LN and BE were taken directly from gauging stations and that at BN was calculated by subtracting former with the later and by regarding the travelling time of the flood wave between BN and LN. Event loads of SPM were calculated as follows: SPM values were interpolated and multiplied with corresponding discharge and the resulting transport rate was integrated over the entire flood period.

RESULTS AND DISCUSSION

DECEMBER 1993 STORM EVENT

In Dec. 1993, heavy, continuous rainfall in the black forest region caused the highest discharge in Enz River since the beginning of its discharge record in 1922. Enz discharge was about two third of Neckar inflow at the confluence, but the peak was delayed by about 6 hours from that of Neckar River (Fig. 2a).

In Fig. 2b, curves of SPM concentration mostly resemble the hydrographs at the corresponding station. Maximum concentrations of SPM appeared about 2 hours ahead of the discharge peak. The peak concentration at BE of about 1200g/m³ is slightly smaller than that at BN ($SPM_{BN,max} = 1325g/m^3$). Much higher concentrations of SPM, up to 1666g/m³, were measured at LN station.

Tab. 1: Event loads of SPM.

| Station | SPM load [*10³ tons] | |
	Dec. 1993	Apr. 1994
BE	53	8.15
BN	107	192
KN	n.d.	204
LN	196	228

n.d.: not determined

SPM loads of Dec 1993 storm event are listed in Tab. 1. The total mass of SPM flowing into the considered river section was 160,000 tons where the contribution of Neckar River being twice as much as Enz River. At Lauffen, SPM event load was measured as 196,000 tons. For the considered river-lock, difference between efflux and influx over the flood period comprises 36,000 tons of SPM.

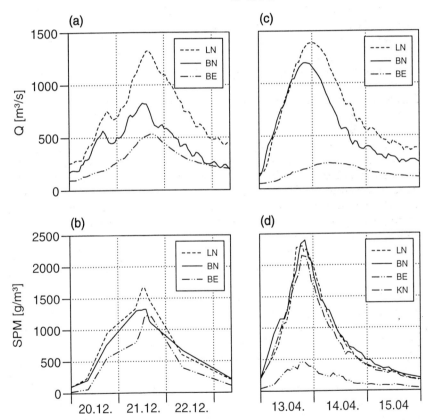

Fig. 2: Discharge and concentration of SPM during Dec. 1993 (a, b) and Apr. 1994 (c, d) storm events.

APRIL 1994 STORM EVENT

In Apr. 1994 extensive precipitation occurred in the Swabian mountains and in the upper Neckar area and caused a rapid increase of Neckar River discharge (Fig. 2c). On April 13th, it reached a maximum of 1200m³/s at BN, which was much higher than peak discharge of 823m³/s in Dec. 1993. In contrast, maximum discharge of Enz River (243m³/s) didn't reach half of the preceeding flood peak. The tributary's peak discharge developed at about 10 hours later than that of Neckar River.

In the Enz River, concentration of SPM was significantly low, compared to that in the Neckar River (Fig. 2d). SPM curves at stations BN, KN and LN depict a similar course. Due to mixing of Neckar with Enz water in the upper section of the river-lock, SPM values at KN are smaller than those at BN. Higher concentration of SPM at LN than at KN indicate resuspension of sediment in the lower reach of the river-lock.

At Neckar River, peak values of SPM and discharge coincide, whereas at Enz River, SPM concentration reaches its maximum about 15 hours earlier than discharge does (Fig. 2c, 2d). This implies that sediment supply in the catchment of Enz River was almost exhausted due to the Dec. 1993 flood. On the contrary, SPM concentration in Neckar River is more limited by the hydraulic conditions than by sediment supply from channel bed or external sources.

Mass balance for Apr. 1994 event listed in Tab. 1 shows that mass of SPM flowing into the river-lock is almost entirely formed by the Neckar River (192,000 tons); Enz River contributes only 8,150 tons. The sediment load of 204,000 tons determined at KN corresponds very well to the total inflowing load of about 200,000 tons from BN and BE. The amount of SPM leaving the river-lock at Lauffen encompasses 228,000 tons, which exceeds load at KN by 24,000 tons. In the lower section of the river-lock between KN and LN, erosion is predominant over deposition.

RELIABILITY OF SPM DATA

Collection of representative samples in fluvial cross sections may be affected by non-uniform distribution of SPM in vertical or lateral direction (Ongley et al., 1990; Horowitz et al., 1990). Mean grain-size of the suspended particles in the regarded section was found to be of diameter ranging from 10 to 20μm at flood events (data not presented here) and thus, according to "diffusional theory" (Ippen, in Rouse, 1937), almost complete vertical mixing can be anticipated. Lateral variation of SPM concentration was determined from near-surface samples collected at several positions of the river cross-sections at stations BE, BN and LN.

In Fig. 3, concentration of SPM is plotted against relative position of sampling points in the cross-section, which is expressed by the ratio of distance from left bank (b) to the channel width (B). At BE, complete mixing was observed over the cross-section, whereas transversal gradient in SPM concentration was found at LN and less significantly at BN. Higher concentrations of SPM at left bank positions of BN and LN are obviously due to non-uniform erosion of channel sediments in the Neckar River. In the backwater sections at Besigheim and Lauffen, navigation traffic runs along the right bank. Turbulence by its movement prevents fine-grained particles from settling at this side. At left bank side, higher sedimentation rate during low discharge periods leads to greater supply of sediments which can be eroded by hydraulic bottom shear at high discharge.

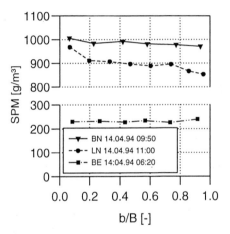

Fig. 3: Cross-sectional profiles of SPM concentration.

At station LN, centroid of flow is located between b/B=0.2 and b/B=0.55 (Fig. 3) as other part of the cross-section is flooded only at high discharge. At the sampling position used throughout the storm events, concentration of SPM deviates by only 0.7% from flow-weighted cross-sectional mean (Tab. 2). Thus, in spite of the lateral gradient in SPM profile, the sampling point at LN proved to be representative. Same can be inferred to BE and BN sampling stations, where deviation is about 1%, only.

Tab. 2: Deviation of single-point SPM concentration from flow-weighted cross-sectional averages on 14.04.1994.

Station	B [m]	Sampling point b/B [-]	Sampling point SPM [g/m³]	Cross section SPM [g/m³]	Deviation [%]
BE	65	0.50	228 at b/B=0.43	231	1.6
			234 at b/B=0.57		1.3
BN	78	0.42	992	984	0.7
LN	162	0.33	907	901	0.7

CONCLUSIONS

The principle of mass continuity is applied to suspended particulate matter in the lock-controlled river section. With respect to SPM concentrations used in the mass balances, sampling points showed only about 1% deviation from the flow-weighted value of the corresponding fluvial cross-section. An excess of outflowing over inflowing load of SPM was determined for both, the Dec. 1993 storm event (32,000 tons) and the Apr. 1994 flood (24,000 tons). These amounts express a dominance of erosion over deposition of sediments in the river-lock at high discharge. As measured in the second flood event, net erosion occurred in the lower section of the lock, where fine-grained sediments deposit under backwater conditions at low discharge. At this section, morphological changes of the channel bed due to other flood events were also observed through echo-sounding surveys carried out at a spacing of 100m (Kern and Westrich, 1995b).

ACKNOWLEDGEMENTS

Funding of this research was provided by Projekt Wasser-Abfall-Boden (PWAB). My thanks go to V. Schürlein and R. Chakraborti for their contribution and practical support of the project.

REFERENCES

Förstner, U. and Wittmann, G.T.W. (1981): Metal pollution in the aquatic environment. Springer, New York.

Horowitz, A.J.; Rinella, F.A.; Lamothe, P.; Miller, T.L.; Edwards, Th.K.; Roche, R.L. and Rickert, D.A. (1990): Variations in suspended sediment and associated trace element concentrations in selected riverine cross sections. *Environmental Science and Technology* **24**, 1313-1320.

Kern, U. and Westrich, B. (1995a): Sediment contamination by heavy metals in a lock-regulated section of the river Neckar. Submitted to: *Marine and Freshwater Research*.

Kern, U. and Westrich, B. (1995b): Sediment dynamics in a lock-regulated section of the Neckar River. Submitted to: *Archiv für Hydrobiologie, Beihefte Ergebnisse der Limnologie*.

Ongley, E.D.; Yuzyk, T.R. and Krishnappan, B.G. (1990): Vertical and lateral distribution of fine-grained particulates in prairie and cordilleran rivers: sampling implications for water quality programs. *Water Research* **24** (3), 303-312.

Rouse, H. (1937): Modern concepts of the mechanics of fluid turbulence. *Trans. ASCE* **102**, 463-543.

OBSTRUCTION LIMITS BENEATH THE SUBMERGED HORIZONTAL PLATE WAVE FILTER

SILKE KLIETSCH
Student, University of Wuppertal
Pauluskirchstr. 7, 42285 Wuppertal, Germany

The physical processes near a horizontal submerged plate applied beneath waves are very complicated. A flow will be created beneath the plate if several conditions are respected. The aim of the study presented was to determine some of these conditions. It is shown that the physical behaviour of the plate wave filter will be destroyed if the obstruction beneath the plate is too strong. A possible construction beneath the plate has to be designed very carefully if the original behaviour should be kept.

REASONS FOR THE INVESTIGATION

A rigid horizontal submerged plate used for the wave height reduction has several advantages compared to conventional breakwaters. Its operation as a wave filter allows a profitable solution which also makes an ecologically desirable exchange of water between the open sea and the area to be protected possible. Critical comparisons of the literature were done by Graw [1994 and 1993-1].

By varying all parameters the hydrodynamical system close to the plate reacted very sensitively upon changes, so that an enormous program of measurements should be necessary for an analytical description.

Many authors do not mention the flow near the plate, but for example Guevel et al. [1985] give detailed descriptions that they observe just a little motion forewards and backwards beneath the plate. Dick [1968] was the first author refering to a directed flow beneath the plate. Further experiments performed by Graw [1993-2 and 1993-3] led to the idea to use the flow created beneath the plate for the development of a wave power converter. For the realization of such a combined breakwater and power station more knowledge about this flow is necessary, therefore the presented study was undertaken.

EXPERIMENTS

For the experiments one plate length and one submergence depth were chosen according to potential constructions and in accordance with earlier experiments in which a strong flow beneath the plate was observed. It was necessary to conduct the parameter variation L/l gradually with very small changes because an extraordinary discontinuous behaviour had to be described. The variation of the wave length was limited by the wave generator. Measurements with the initial wave heights $H_i = 0.02$ m and $H_i = 0.03$ m were performed.

The experiments conducted to optimize the wave height reduction and the intensity of the pulsating flow by means of a partial reduction of the cross section under the plate are based on the continuity equation $Q = v\,A$. It is theoretically possible to achieve infinitely high velocities for a very small cross section. But there are actually further factors limiting the possible velocity, for example friction losses.

The investigation of the effects of the kind and degree of obstruction under the plate on the flow behaviour near the plate was subdivided into three series of measurement.

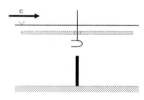

Figure 1: Setup of series 1

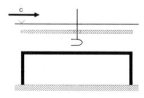

Figure 2: Setup of series 2

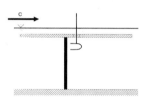

Figure 3: Setup of series 3

- Series 1 - Obstacle with a circular hole: In the first series of measurement the obstacle was created by a screen with a circular hole *(figure 1)*. The screen was placed on four different positions under the plate. The outside measurements of the screen corresponded to the rectangular cross section (780 cm²) under the plate. The circular hole in the upper third of the screen was 7.07 cm², the remaining cross section where the water could flow was (opening ratio =) 0.9 %.
  The reason for this measurement series was to examine the behaviour of the system for very low opening ratios.

- Series 2 - Obstacle by means of a wall: Two vertical walls of different height were placed under the centre of the plate, the opening ratios were 38.5 % and 46.2 % *(figure 2)*.

- Series 3 - Obstacle by means of a caisson: Two caissons of different height were placed under the plate, the opening ratios were 32.7 % and 40.4 % *(figure 3)*.

The reason for the measurement series 2 and 3 was to examine the behaviour of the system for different flow patterns created by the obstacles.

ANALYSIS

The results of the measurements (wave height reduction C_t behind and flow velocity v_x beneath the plate) are shown graphically with respect to the relative wave length L/l of the initial wave.

Series 1 - Obstacle with a circular hole

The C_t-graphs of the measurements with the different positions of the screen are (with respect to the highest accuracy) almost identical *(figure 4)*. The graph of the transmission coefficient of the non-obstructed plate shows a similar, slightly shifted course for $H_i = 0.02$ m; the minimum value, however, is developed considerably stronger. The non-obstructed plate can be called a wave filter while the obstructed plate reduces the wave height of different wave lengths relatively uniform.

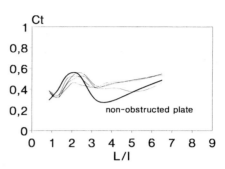

Figure 4: C_t-graphs of series 1, $H_i = 0.02$ m

The graphs of the measurements with $H_i = 0.03$ m differ almost in the whole measurement range compared to the non-obstructed plate. The partial obstruction of the cross section under the plate leads to a reversal of the results: The wave height reduction of long waves is best by non-obstructed plates whereas the wave height reduction of short waves (L/l $\leq$ 2.0) is better if the region below the plate is obstructed. The characteristics of the wave height reduction of a plate under which the cross section is closed up to about 1 % opening ratio is completely different compared to a non-obstructed plate. The new construction can no longer be called a wave filter, it acts more like a porous rubble underwater breakwater.

The measurements conducted with the screen at a position near the middle of the plate showed a distinct enlargement of the flow velocity which occurred under the non-obstructed plate. In the range of the large relative wave lengths (L/l $\geq$ 3.5) the obstruction under the plate led to a velocity value up to seven times higher than the values measured for the non-obstructed plate (from $v_a \approx 5.0$ cm/s to $v_a \approx 35.0$ cm/s). Because this is not a directed but an oscillating flow, for a wave energy power station based on this configuration we would have to use a turbine like the Wells turbine which is suitable for the extraction of energy from reversed cyclic flows. It would use the large amplitude of the velocity in both directions for the production of energy.

Series 2 - Obstacle by means of a wall

The course of the C_t-graphs (opening ratio 38.5 % and 46.2 %) of one wave height show a very good conformity, they do also correspond essentially to the graphs determined without any obstruction under the plate *(figure 5)*. All deviations occurring

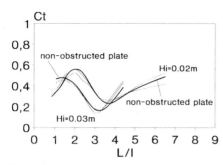

Figure 5: C_t-graphs of series 2

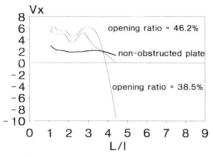

Figure 6: V_x-graphs of series 2, H_i = 0.03 m

are within the scope of the accuracy of the used method and can therefore be regarded as not significant.

Both configurations cause a constantly strong flow around the plate for short relative wave lengths (L/l ≤ 3.0, *figure 6*). This results in good possibilities for the use of energy; but - as for the "normal" plate - the characteristics of the wave height reduction are to a great extent depending on the wave length in this range. It is striking when conducting measurements with this kind of obstruction that the longest wave for the wave height H_i = 0.02 m and opening ratio 38.5 % produces a very strong flow, but in the direction of the wave propagation. This behaviour can also be assumed for the plate with the opening ratio 46.2 %. The physical procedure seems to be changed here.

Series 3 - Obstacle by means of a caisson

Regarding the obstruction under the plate by means of a caisson the C_t-graphs only show a tendentious conformity of the courses of both wave heights. The general difference between the courses is that the measuring data of the greater opening ratio (40.4 %) is of lesser value than the measuring data of the lesser opening ratio (32.7 %).

Concerning to this obstruction the plate with the greater opening ratio causes a better wave height reduction of nearly all wave lengths.

A conformity of the courses of the non-obstructed plate and the obstructed plate can be noticed only in the measuring area of the relative wave length 1.5 ≤ L/l ≤ 3.0. The non-obstructed plate reduces the wave height of longer waves better.

In comparison to the experiments

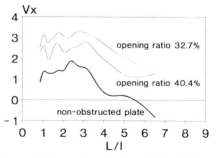

Figure 7: V_x-graphs of series 3, H_i = 0.02 m

with the rigid submerged plate the obstructed plate produces an increased flow velocity. With the wave height $H_i = 0.02$ m the smaller opening ratio (32.7 %) causes an increase of the flow velocity from $v_m \approx 1.5$ cm/s to $v_m \approx 3.0$ cm/s, which is nearly twice as much.

Comparison of series 2 and 3
Summing up one can say that the opening ratio has only little influence on the wave height reduction. The different kinds of obstructions however show a distinct influence - at least regarding large and small relative wave lengths.
It can also be seen that both kinds of obstructions produce a stronger flow than the non-obstructed plate. With the wave height $H_i = 0.02$ m the plate with the caisson obstruction produces a stronger flow than the plate obstructed with the vertical wall.

RESULTS
Opening ratio between 30 % and 50 %
- Concerning the wave height reduction there is no improvement compared to the non-obstructed plate.
- The plate with the caisson obstruction partial causes a worse wave height reduction than the non-obstructed plate. Regarding the vertical wall obstruction there is only a slightly different wave height reduction compared to the non-obstructed plate.
- The courses of the transmission coefficient show distinctive peak values, i. g. the function of the submerged plate as a wave filter is seen also for these kinds of obstructions.
- All used obstructions and opening ratios result in a stronger pulsating flow beneath the plate.
- Regarding the wave height $H_i = 0.02$ m there is a more or less uniform pulsating flow against the direction of wave propagation in the whole measuring area in despite of the different kinds of obstruction. The intensity of the pulsating flow with the wave height $H_i = 0.03$ m shows distinctive peak values.

Opening ratio 1 %
- The wave height reduction features of a plate where the region below the plate is closed as far as approximately 1 % opening ratio do not correspond to the function of the submerged plate as a wave filter.
- The restriction of the cross-sectional area of the flow beneath the plate as far as 1 % opening ratio produces a different flow behaviour (always a pulsating flow).

CONCLUSIONS AND OUTLOOK
The analysis of the influence of an obstruction beneath the submerged horizontal plate on the performance of this construction was only a small part of a larger research project. The main object of the investigation was to examine the physical phenomenon of the pulsating flow beneath the plate. In order to understand all conditions

influencing this phenomenon a comprehensive variation of parameters is necessary, involving an immense number of experiments.

The results presented here supplement the existing knowledge about the plate wave filter respectively the horizontal plate wave power converter, as it was shown for the first time, that it is not possible to obstruct the region below the plate too much without destroying the physical phenomenon of the plate wave filter. It was demonstrated furthermore, that measures to enhance the flow below the plate (caisson instead of plate) influence the wave height reduction. Nevertheless the behaviour of the plate wave filter was still achieved with both types of obstruction with openings of only 1/3 of the cross section beneath the plate.

Ongoing measurement series will be performed in the near future to examine the influence of different shapes of the obstruction below the plate and to determine the maximum obstruction possible without change of the physical behaviour near the plate. As well as the experiments presented, these experiments are necessary for a future realization of the new type of wave power converter.

REFERENCES

Dick, T.M.; *On solid and permeable submerged breakwaters;* 1968; Ph. D. Diss. at Queens University Kingston

Graw, K.-U; *Untersuchungen am Plattenwellenbrecher;* 1994; Mitteilung 7, Lehr- und Forschungsgebiet Wasserbau und Wasserwirtschaft, Bergische Universität Wuppertal

Graw, K.-U.; *The submerged plate wave energy converter (A new type of wave energy device);* 1993; Proc.: Int. Symposium on Ocean Energy Development (ODEC), Muroran, pp. 307-310

Graw, K.-U.; *Shore protection and electricity by submerged plate wave energy converter;* 1993; Proc.: European Wave Energy Symposium, Edinburgh, pp.379-384

Graw, K.-U.; *The submerged plate as a primary wave breaker;* 1993; Proc.: XXV IAHR Congress, Tokyo, pp.38-45

Guevel, P.; Landel, E.; Bouchet, R.; Manzone, J.M.; *Le phénomène d'un mur oscillant et son application pour protéger un site côtier soumis à l'action de la houle;* 1985; Bulletin de l'Association Technique Maritime et Aeronautique, Vol. 85, pp.229-245

Klietsch, S.; *Labormessungen zu den Möglichkeiten der Beeinflussung des Wirkungs-grades des Plattenwellenbrechers durch Veränderung der Plattenumströmung;* 1993; Diplomarbeit, Wasserbau und Wasserwirtschaft, Bergische Universität Wuppertal

HydroInformatic Applications in Real Time Control Strategy Selection

KHONDKER MASOOD-UL-HASSAN and GEOFFREY WILSON
Research Fellows, International Institute for Infrastructural,
Hydraulic and Environmental Engineering, Delft, The Netherlands.

ABSTRACT

Two applications, using techniques associated with artificial intelligence, are presented for control strategy selection in multi-objective water resource systems operated in real time. The first demonstrates the use of an artificial neural network in replicating optimised gate setting patterns in a large flood control scheme. The second shows that a classifier system is capable of rule base learning for the control of an urban drainage system.

INTRODUCTION

Real time control (RTC) of water resources involves dynamically operating regulators (pumps, gates, weirs, valves, etc.) based on remotely transmitted measurements (rain, discharge, water level, etc.) during the event to improve system performance. Real time control systems must be able to monitor and respond to processes in the real world in the face of time constraints.

A control strategy is the time sequence of regulator settings or set-points in a dynamically operated system. Control strategies are associated with one or more operational objectives. The operational objectives of a large flood control scheme, as in the first application, are *i)* in cases of large floods, to control floodplain usage and mitigate downstream flood damage, and *ii)* in smaller floods, to maintain optimum floodplain water levels to achieve ideal cropping conditions. In an urban drainage system the objectives are typically *i)* to reduce environmental damage due to combined sewer overflow (CSO), *ii)* to reduce surface flooding, *iii)* to equalise treatment plant inflows and *iv)* to reduce operational and maintenance costs. These objectives introduce multiple - and conflicting -criteria decision-making. One way of solving such a decision-making process is to quantify each objective in a relative sense with a cost function. Costs are assigned to selected flow variables in such a way that to minimise the cost function over some time interval is to achieve *optimal* control. Most non-linear mathematical optimisation routines require a large number of iterations to converge to a quasi-optimal solution. In problems of the present type, non-linear optimisation techniques require a large number of simulations to compute cost gradients. These techniques are too computationally demanding and too slow to be effectively employed in real time.

This paper presents two new applications of existing technologies where cost function minimisation is used as a quantification of optimal control and where control behaviour is learned from experience. Implementation response is near real time. It appears that both technologies are suitable for integrated management of basin-wide resources operated under RTC.

APPLICATION OF ARTIFICIAL NEURAL NETWORKS
Artificial Neural Networks (ANN) have been widely applied to pattern recognition problems that resemble the problem of interest in the present context. An ANN is one of a number of information processing technologies capable of machine learning. The type of network chosen was the fully connected, feed forward, error back-propagation algorithm which uses the generalised delta rule for updating weights. This algorithm is attributed to Rumelhart et al (3). Machine learning of the supervised type was employed in this study by attempting to duplicate the behaviour contained in a data set (input:output). During training, the data are presented repeatedly to the ANN which improves an input to output mapping by updating connection weights. The ANN thus learns an appropriate mapping from experience. During verification a trained network will respond to new data (input), that were never presented during the training phase by providing an appropriate output pattern.

An application is presented in which ANN modelling has been used to replicate optimised gate settings obtained from a numerical optimisation routine. The application is to a real catchment in the North Central part of Bangladesh. The objective is to maintain a desired water level at a control location by controlling 5 local hydraulic structures. These structures are referred to as Loh1, Binn, Dha2, Dha3 and Dha4. Due to the topology and hydraulics of the area, it has been seen that the water level at the control location is representative of the entire study area.

The data (input:output) were obtained, every 15 minutes, from MIKE11 (DHI's 1-D hydrodynamic model) simulations coupled to a numerical optimiser based on the gradient search method. The input were upstream and downstream water levels, flows and current gate setting at the 5 gate locations, and either water levels or flows at 6 regional locations. The output were desired gate settings for a time horizon. Three time horizons were investigated, 15 minutes, 60 minutes and 1 day. Approximately 70 percent, chosen randomly, were used for training and the remaining 30 percent were used for validating the network.

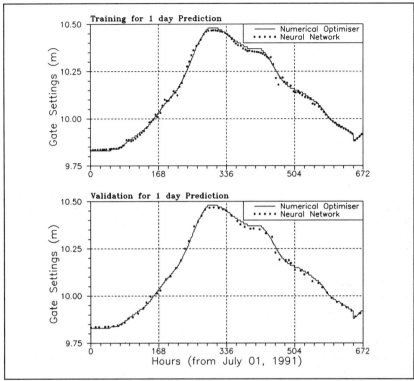

Fig. 1:   1 day prediction results for Dha4.

Prediction Horizon	Loh1	Binn	Dha2	Dha3	Dha4
15 min (Training)	0.999	0.999	0.999	0.999	0.998
60 min (Training)	0.998	0.999	0.999	0.999	0.988
1 day (Training)	0.996	0.998	0.997	0.998	0.983
15 min (Validation)	0.999	0.999	0.999	0.999	0.998
60 min (Validation)	0.998	0.998	0.999	0.999	0.991
1 day (Validation)	0.996	0.998	0.997	0.997	0.978

Table 1:   The coefficients of determination, r, for training and validation.

Results reflecting the high degree of performance, even up to the 1 day prediction horizon, are presented. Fig. 1 presents typical training and validation results for desired gate settings 1 day in advance. Table 1, shows the coefficient of determination, r, between the learned and desired results. The maximum difference between optimised gate settings and ANN gate settings is in the order of 5 cm. The coefficient of determination is very close to unity. For practical purposes the gate setting predictions for the three time horizons (15 and 60 minutes and 1 day) are completely replicated.

APPLICATION OF CLASSIFIER SYSTEMS
Learning classifier systems (1,2), a technique connecting artificial intelligence and genetics, have been applied in this study to the problem of control strategy selection for real-time control of urban drainage systems. A classifier system is a rule-based system that guides behaviour in arbitrary environments where generalisation over states is possible. A learning classifier system learns appropriate behaviour from experience, is adaptive and escapes from the brittleness inherent in traditional rule based systems like expert systems.

Classifiers are finite length, L, ternary {0,1,#} coded CONDITION : ACTION rules. The indifference symbol, #, allows generalisation over states within an accuracy criterion. A Q-type reinforcement learning algorithm (4,6) is the basic learning mechanism. Rule induction is provided by a variation of the traditional Genetic Algorithm (GA). The system objectives are transformed to a cost function. Each rule maintains a prediction of the minimum discounted sum of future costs associated with performing its ACTION part. By updating the cost prediction based on experience, a complete, generalised and accurate mapping from states and actions to cost predictions, $S \times A \Rightarrow C$, is learned.

In the performance algorithm, sensors provide environmental data which are categorised and coded to a binary {0,1} message of finite length, L. Rules are said to match when each bit in the rule CONDITION part matches with the corresponding bit in the environmental message. The # matches unconditionally. Rules that match the environmental message provide predictions on the future cost associated with each ACTION. In the event of encountering an empty match set during learning, classifiers are covered (generated) for each possible action. During learning, action decisions can be chosen stochastically, corresponding to exploratory behaviour, or deterministically, by simply exploiting the minimum cost prediction. Once learned, actions are chosen deterministically only. Chosen actions are decoded from their binary categorisation and effected through the environmental interface.

In the learning phase of the present application hydrological and hydraulic simulations were performed with DHI's MOUSE package. At regular intervals during each simulation the classifier system performed an action (regulator setting) decision. Prior to any new action decision, the system evaluated the cost associated with the previous action and simply added this cost to the smallest discounted cost prediction of each action in the new action set. This Q-like value was used to update the cost prediction of the previous action decision by applying the Widrow-Hoff (5) updating technique. Classifiers also maintain an estimate of prediction error and fitness. These were also updated.

An induction algorithm is a background process which uses a GA to generate new rules. This is done by a recombination of existing highly fit classifiers which replace low performance classifiers. The new rules are then subject to performance and updating pressures to evaluate their role in the overall system. The classifier system GA differs from the traditional GA by considering overlapping generations, partial replacement and ternary mutation. The classifier GA operates in niches provided by the match sets, instead of panmictically and classifier fitness is measured by the accuracy of prediction. In this way there is a natural tendency towards general, but accurate, classifiers.

The learning classifier system developed in this project has been applied to the control of a large, 750,000 population equivalent, sewer catchment in Göteborg, Sweden. The objective was to reduce CSO volume and surface flooding depth-duration and to equalise flows to the treatment plant. These objectives were translated into a cost

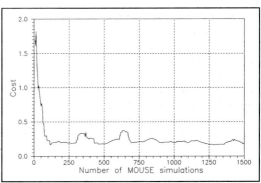

Fig. 2: Typical classifier system learning

function. Water-levels, flows and inflow volume, each categorised into 2, 4 or 8 categories, relating to 1, 2 or 3 bits, were included in the 20 bit environmental sensory message. The action decision was the pump rate to the treatment plant which was categorised into 16 discrete pump rates (4 bits).

Fig. 2 presents the learning associated with a typical 5 hour rainstorm where action decisions were made every 15 minutes. The classifier system has learned to operate the pump to quasi-minimise the cost function evaluation for this event. Once a range of rainstorm events have been learned it is expected that potential benefits may be gained in implementing the classifier system as a decision support tool for real time control of the Göteborg drainage system.

DISCUSSION
Two applications of different technologies have been presented which circumvents the time constraint problem in real time control of basin-wide water resources.

Anns, when implemented correctly, are tools that can be extremely accurate in forecasting. The results presented cover one hydraulic event. Proper generalisation requires training the network on a wide range of events. It appears that the ANN could either completely replace the optimiser at the field level or could complement the optimiser by generating initial estimates, which is expected to speed up the optimising process to allow implementation in real time.

The learning classifier technique employs a fully dynamic hydraulic model and learns based on experience. It is therefore less constrained by simplifications inherent in traditional techniques and presents a move towards a more general tool for control strategy selection. Learning classifier systems are capable of operating a multi-objective urban drainage system. Moreover, the work provides a new research direction for rule-based systems in real-time control of urban drainage systems where traditional rule-based systems are too brittle and where generalisation over states is desirable.

ACKNOWLEDGEMENTS
The work has been carried out at the Danish Hydraulic Institute (DHI) in partial fulfilment of an HydroInformatics M.Sc degree of the IHE, Delft, The Netherlands. The authors wish to acknowledge the advice of Prof. M.B. Abbott, A.W. Minns, L. Yde, V. Babović and S.W. Wilson and the financial support of the Danish Hydraulic Institute.

REFERENCES
1. Goldberg, D.E.(1989) Genetic Algorithms in Search, Optimisation and Machine Learning. Reading, MA: Addison-Wesley.
2. Holland, J.H.(1975) Adaptation in Natural and Artificial Systems. Ann Arbor: The University of Michigan Press.
3. Rumelhart, D.E. and McClelland, J.L.(1988) Parallel Distributed Processing MIT Press, Cambridge.
4. Watkins, C.(1989) Learning from Delayed Rewards. Ph.D. Dissertation, Cambridge University.
5. Widrow, B., & Hoff, M.E.(1960) Adaptive Switching Circuits, New York:IRE, WESCR, pp. 96-104.
6. Wilson, S.W. Classifier Systems Based on Accuracy. Submitted to Evolutionary Computation (1995).

The Effect of Berm Inclination on Flow Structures in Doubly Meandering Compound Channels

C. NAISH & C. A. M. E. WILSON
Department of Civil Engineering, University of Bristol, UK

INTRODUCTION

The construction of a major trunk road along the River Blackwater valley in Hampshire has resulted in the need to relocate a length of the River Blackwater. This gave the National Rivers Authority an opportunity to construct and study the hydraulic performance of an environmentally acceptable, doubly meandering, two-stage channel.

The test reach of the relocated River Blackwater channel has a meandering inner channel with medium sinuosity and a wider upper channel (the floodplain or berm) of low sinuosity. Figure 1 details the planform of the test reach, Figure 2 details a typical cross-section.

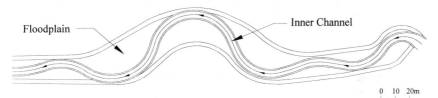

Figure 1 - Planform of the inner channel and floodplains (prototype scale)

Figure 2 - Typical channel cross-section (prototype scale)

The prototype was constructed with berms inclined at 1/30 but model tests at 1/25 and 1/5 scales have been carried out with horizontal and inclined berms. Sellin, Lambert & Werner (1993) carried out the 1/5 scale model study and they found that the use of a berm crossfall increases the conveyance capacity at lower overbank depths as the flow is guided to follow the direction of flow in the main channel. At high overbank depths this effect becomes insignificant and the discharge compared to that for floodplains with no inclination is less because the flow area is smaller.

The adoption of mild crossfalls on the floodplain is recommended for the following reasons :-

• The cross-section that results will give a low berm area immediately adjacent to the main channel which will encourage the development of a marsh habitat.
• The portion remote from the main channel will be correspondingly high and drier and will provided better access for maintenance machinery.

Wherever possible therefore, a berm crossfall of about 1 in 30 should be retained and additional cross-sectional area if required can be provided by increasing the channel overall width.

APPARATUS AND EXPERIMENTAL PROCEDURE
The undistorted 1/5 scale model of the River Blackwater was constructed in the 56m long, 10m wide SERC FCF. Figure 3 shows the planform of the model and the location of the cross sections at which flow velocity and direction measurements were taken. Instrument details for the FCF are to be found in Knight and Sellin (1987).

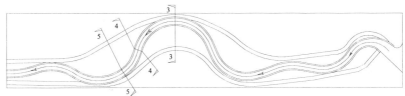

Figure 3 - Location of cross section positions

To investigate the structure of the flow in the model an extensive series of flow direction and velocity measurements were undertaken at various flow depths at Sections 3, 4 and 5 using a miniature vane for directional measurement and a miniature propeller, pointed in the direction of the flow, for velocity measurement. Both were mounted on a fully automated instrument carriage.

Figure 4 - 1/5 scale River Blackwater model with the automated instrument carriage surveying Section 4.

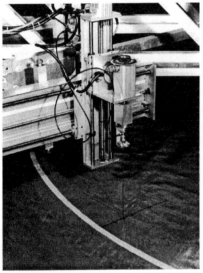

Figure 5
Instrument carriage, miniature vane
and propeller details.

The following velocity distribution diagrams were constructed from the 1/5 scale model data by resolving the velocity data parallel to the local main channel direction for the longitudinal velocity contour diagrams and resolved perpendicularly to this direction for the transverse velocity profiles. All sections are drawn looking downstream.

HORIZONTAL BERM LONGITUDINAL VELOCITY
CONTOURS & TRANSVERSE VELOCITY PROFILES

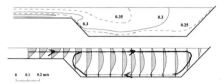

At Section 3, with a high (187mm) overbank flow depth, the main channel roughened with 8mm gravel and the horizontal floodplain roughened with 13mm gravel, the longitudinal velocity contours indicate a main channel longitudinal velocity range of between 0.25 and 0.35 m/s. The maximum velocity core is positioned on the inside of the main channel bend. The floodplain longitudinal velocity range is lower at between 0.2 and 0.35 m/s. The transverse velocity profile suggests a large, single, clockwise, floodplain driven, main channel cell.

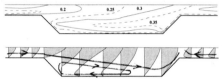

At Section 4, it can be seen that the main channel longitudinal velocity range has been increased to between 0.2 and 0.35 m/s with the maximum velocity core having moved to the outside of the bend. The longitudinal floodplain velocity has increased to match the local main channel longitudinal velocity, with the inside floodplain longitudinal velocity slower than the outside floodplain longitudinal velocity. The transverse velocities have increased with travelling towards the floodplain/main channel/floodplain crossover region. A clockwise main channel cell still exists but has been squashed by the inside floodplain flow crossing the main channel. The outside bend floodplain is trying to discharge into the main channel but the strong main channel transverse currents prevent this and the flow is forced downstream.

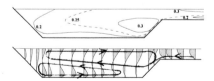

At Section 5, the main channel longitudinal velocity range has reduced to between 0.2 and 0.3 m/s with the maximum velocity core remaining on the inside and the longitudinal floodplain velocities have been reduced. The transverse velocity profiles show that the main channel cell seen at Section 4 remains, but with its flow direction reversed. A second, clockwise rotating, upper main channel transverse velocity cell has developed which is now taking water off the floodplain.

INCLINED BERM LONGITUDINAL VELOCITY
CONTOURS & TRANSVERSE VELOCITY PROFILES

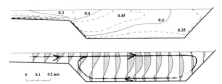 At Section 3, with a high (187mm) overbank flow depth, the main channel roughened with 8mm gravel and the inclined floodplain roughened with 13mm gravel, the longitudinal velocity contours indicate that the main channel longitudinal velocities are 0.1 m/s higher, with the same velocity distribution, when compared with the horizontal berm case. The floodplain longitudinal velocities have remained relatively unchanged. The transverse velocity profile has not been affected by the berm inclination, but the transverse velocities have increased, and the large, single, clockwise rotating, floodplain driven, main channel cell remains.

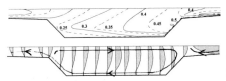 At Section 4, the main channel velocity range has increased to between 0.25 and 0.5 m/s, compared with the 0.2 to 0.35 m/s range experienced with the horizontal berm condition. Again, the floodplain velocities remain unchanged. The main channel longitudinal flow structure remains as with the horizontal berm condition, with the longitudinal velocity core positioned on the outside of the main channel bend. The effect of the berm inclination on the transverse velocity profiles is shown clearly here, as the cross channel floodplain flow is now prevented, leaving a single, clockwise rotating, transverse velocity cell to occupy the whole of the main channel.

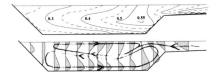 At Section 5, the main channel longitudinal velocity range has increased from between 0.2 and 0.3 m/s with a horizontal berm to between 0.25 and 0.55 m/s, and the inclined floodplain longitudinal velocities are 0.1 m/s higher. The upper main channel transverse velocity cell has reduced in size laterally and has been pushed towards the outside of the main channel bend, with the lower, anticlockwise rotating transverse velocity cell growing in size and strength.

CONCLUSIONS

Inclining the berms increases the longitudinal velocities in the main channel, with the high velocity core retaining its position on the inside of the main channel bend.

Floodplain longitudinal velocities are unaffected by the berm inclination.

The effect of berm inclination is most clearly marked by the change in the transverse velocity cell structure. At the bend apexes, for both horizontal and inclined berms, the main channel cell structure is driven by floodplain flow entering the main channel. At the floodplain/main channel/floodplain cross-over region, the cross channel flow experienced with horizontal berms is prevented by the berm inclination. The main channel transverse velocity cell, squashed by the cross channel horizontal berm flow, expands ver⁺ically to cover the whole main channel cross section.

The overall effect of inclining the berms is to constrain the flow to a course more closely aligned with that of the main channel, therefore reducing the floodplain/main channel flow interaction.

ACKNOWLEDGEMENTS

The authors with to acknowledge the financial support of SERC and the NRA in this research and the work carried out on the 1/5 scale model by Dr M Lambert of the University of Newcastle, Australia.

REFERENCES

Sellin R H J, Lambert M F and Werner M G F (1993), Large scale model investigation of a two stage channel, NRA R&D note 366, September 1993.

Knight D W & Sellin R H J (1987), The SERC Flood Channel Facility, Journal of the IWEM, Vol 1, No 2, pp198-204, October 1987.

OPTIMIZATION OF CALCULATION FOR
NAVIER-STOKES EQUATION
IN SPECIAL COORDINATE SYSTEM   BASED ON
ANALYSES OF FLOW'S
GEOMETRICAL PARAMETERS

RAZUMOVSKY E.M.
Postgraduate student, Saratov State Technical University.
Saratov. Russia.

ABSTRACT

The aim of this reasearch is the optimization of expression for 'viscosity terms' in Navier-Stokes equations without sufficient accuracy losses. Navier-Stokes equations are presented in special nonorthogonal curvilinear coordinate system. This report presents the results of mathematical experiment. These results once obtained can essentcially minimize the occupation of computer resourses in Navier-Stokes equation calculation process.

INTRODUCTION

Viscous terms in Navier-Stokes equations in special coordinate system directly contain geometrical and kinematical parameters of flow such as radii of streamline curvature, speed of fluid, angles of streamline turn etc. This method is useful for afterward analyses of geometrical parameters influence on total friction losses. Navier-Stokes equations expression is rather sophisticated that creates some difficulties for computer calculation. Author tries to evaluate the rate of influence on the whole 'viscous term' for every geometrical parameter. This method enables creating more simple expression for 'viscous term' in particular practical tasks.

GENERAL EQUATIONS
The tensor view for Navier-Stokes equations is:

$$-grad\left(Pm + \int\frac{dp}{\rho}_f\right) + \nu\Delta\vec{U} = \frac{d\vec{U}}{dt} \quad (1).$$

Whear $\nu\Delta\vec{U}$ -is 'viscous term'.The expressions for 'viscous terms' in special coordinate system may be written as follows:

$$T_l = \frac{1}{Re}\{\frac{1}{\cos^2\beta}(\frac{\partial^2 Fr}{\partial l^2} + \frac{\partial^2 Fr}{\partial z^{*2}}) + \frac{\partial^2 Fr}{\partial y_1^2} - \frac{\partial^2 Fr}{\partial l\partial z^*}\frac{2tg\beta}{\cos\beta} +$$

$$+\frac{\partial Fr}{\partial l}(\frac{tg\beta}{R\cos^2\beta} + \frac{1}{\rho_{yl}\cos\beta}) - \frac{\partial Fr}{\partial y_1}\frac{1}{\rho} - \frac{\partial Fr}{\partial z^*}(\frac{1}{R\cos^3\beta} + \frac{tg\beta}{\rho_{yl}}) +$$

$$+Fr[\frac{2\sin\beta\cos\beta}{\rho}\frac{\partial\alpha}{\partial z^*} - \frac{1}{R^2\cos^2\beta} - \cos^2\beta(\frac{1}{\rho^2} + \frac{1}{\rho_{yl}^2}) - \quad (2);$$

$$-(\frac{\partial\alpha}{\partial z^*})^2 - (\frac{\partial\beta}{\partial y_1})^2 - \frac{1}{\cos^2\beta}*(\frac{\partial\beta}{\partial z^*})^2]\}$$

$$T_{yl} = \frac{1}{Re}\{2[\frac{\partial Fr}{\partial l}(\frac{1}{\rho} - tg\beta\frac{\partial\alpha}{\partial z^*}) + \frac{\partial Fr}{\partial y_1}\frac{\cos\beta}{\rho_{yl}} + \frac{\partial Fr}{\partial z^*}(\frac{1}{\cos\beta}\frac{\partial\alpha}{\partial z^*} -$$

$$-\frac{\sin\beta}{\rho})] + Fr[\frac{2\sin\beta}{\rho^2}\frac{\partial\rho}{\partial z^*} - \frac{1}{\rho^2}\frac{\partial\rho}{\partial l} - \frac{\cos\beta}{\rho_{yl}^2}\frac{\partial\rho_{yl}}{\partial y_1} - \frac{2tg\beta}{R\rho} + \quad (3);$$

$$+\frac{4tg\beta\sin\beta}{\rho}\frac{\partial\beta}{\partial z^*} - \frac{\sin\beta}{\rho_{...}}(2\frac{\partial\beta}{\partial y_1} + \frac{\partial\alpha}{\partial z^*}) + \frac{tg^2\beta - 1}{R}\frac{\partial\alpha}{\partial z^*}]\}$$

$$T_{z'} = \frac{1}{Re} \left\{ \left(\frac{\partial^2 Fr}{\partial l^2} + \frac{\partial^2 Fr}{\partial z^{*2}} \right) \frac{tg\beta}{cos\,\beta} + \frac{\partial^2 Fr}{\partial y_1} sin\,\beta - \frac{\partial^2 Fr}{\partial l \partial z^*} 2\,tg^2\beta + \right.$$

$$+ \frac{\partial Fr}{\partial l} \left(\frac{cos^2\beta + 1}{R\,cos^3\beta} + \frac{tg\beta}{\rho_{y1}} - 2\,tg\beta \frac{\partial\beta}{\partial z^*} \right) + \frac{\partial Fr}{\partial y_1} \left(2\,cos\,\beta \frac{\partial\beta}{\partial y_1} - \frac{sin\,\beta}{\rho} \right) +$$

$$+ \frac{\partial Fr}{\partial z^*} \left(\frac{2}{cos\,\beta} \frac{\partial\beta}{\partial z^*} - \frac{tg\beta\,sin\,\beta}{\rho_{y1}} - \frac{tg\beta(2\,cos^2\beta + 1)}{R\,cos^3\beta} \right) + Fr \left[\frac{2\,tg\beta}{R^2} \frac{\partial R}{\partial z^*} - \right. \quad (4);$$

$$- \frac{1}{R^2\,cos\,\beta} \frac{\partial R}{\partial l} + \frac{1}{R\rho_{y1}} + \frac{tg^2\beta - 1}{R} \frac{\partial\beta}{\partial z^*} - \frac{sin\,\beta}{\rho_{y1}} \frac{\partial\beta}{\partial z^*} - \frac{cos\,\beta}{\rho} \frac{\partial\beta}{\partial y_1} +$$

$$+ cos\,\beta \frac{\partial^2\beta}{\partial y_1^2} + \frac{1}{cos\,\beta} \frac{\partial^2\beta}{\partial z^{*2}} - sin\,\beta \left(\frac{\partial\beta}{\partial y_1} \right)^2 - \frac{tg\beta}{cos\,\beta} \left(\frac{\partial\beta}{\partial z^*} \right)^2 \right] \right\}$$

These expressions may be written otherwise :

$$\Pi = f_0 + \frac{\partial U}{\partial l} f_l + \frac{\partial U}{\partial y_1} f_{y1} + \frac{\partial U}{\partial z^*} f_z + U f_U \quad (5);$$

$$T_{y1} = g_0 + \frac{\partial U}{\partial l} g_l + \frac{\partial U}{\partial y_1} g_{y1} + \frac{\partial U}{\partial z^*} g_z + U g_U \quad (6);$$

$$T_{z^*} = h_0 + \frac{\partial U}{\partial l} h_l + \frac{\partial U}{\partial y_1} h_{y1} + \frac{\partial U}{\partial z^*} h_z + U h_U \quad (7).$$

The considerable geometrical parameters and special coordinate system are presented in figure 1:

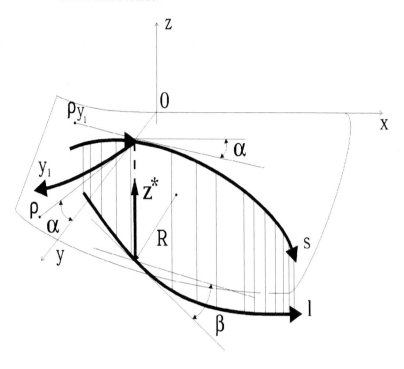

Figure 1: special coordinate system and parameters of
streamline's curvature.

ANALYSES

Using expressions 5,6,7 we can evaluate the influence rate of every
geometrical parameter over each coefficient fi,gi,hi i=0,l,y1,z*,u. Obtained
result makes it possible to evaluate the significance of each geometrical
parameter for various cases of initial data.For example from (2) follow that
f_u depends upon 8 different parameters. It is difficult supperate influence
of some parameter and evaluate its rate without full calculation for 'viscous
term'.

INITIAL DATA

The numerical results are already obtained for f_u in the following area of geometrical parameters:

α	-60°	-45°	-30°	-15°	0°	15°	30°	45°	60°
β	-60°	-45°	-30°	-15°	0°	15°	30°	45°	60°
$\dfrac{1}{R}$	-4	-3	-2	-1	0	1	2	3	4
$\dfrac{1}{\rho}$	-4	-3	-2	-1	0	1	2	3	4
$\dfrac{1}{\rho_{r1}}$	-4	-3	-2	-1	0	1	2	3	4
$\dfrac{\partial\alpha}{\partial z.}$	-4	-3	-2	-1	0	1	2	3	4
$\dfrac{\partial\beta}{\partial y}$	-4	-3	-2	-1	0	1	2	3	4
$\dfrac{\partial\beta}{\partial z}$	-4	-3	-2	-1	0	1	2	3	4

RESULTS

Application of special methematical methods makes it possible to highlight to some extent for every geometrical parameter in which its influence rate in total 'viscous term' is less then 5%. This result may increase essentcially the speed of calculations based on the initial flow geometry data and make the calculation procedure more simple without acuracy losses. In the case of f_u we can exclude parameter $\dfrac{\partial\alpha}{\partial z.}$ from calculations in follows area of intial data:

$\alpha=60^o$; $\beta=60^o$;

$\dfrac{1}{R}$ - from -4 to 4; $\dfrac{1}{\rho}$ -from -4 to 4; $\dfrac{1}{\rho_{r1}}$ -from -4 to 4;

$\dfrac{\partial\alpha}{\partial z.}$ from 2 to 4; $\dfrac{\partial\beta}{\partial y_1}$ -from -4 to 4; $\dfrac{\partial\beta}{\partial z}$ -from -4 to 4;

CONCLUSION

The main output of this research is the creation of simpler expressions for coefficients f,g,h in various kinds of flow geometry. This method makes the procedure of calculation more fast and simple without significant acuracy losses.

ACKNOWLEDGMENTS

Special thanks to professor Lev I. Vysotsky who is the initiator of this research.

NOTATIONS

α - angle of streemline-turn in horizontal direction;

β - angle of streemline-turn in vertical direction;

Fr - Froude number;

l - streemline;

P - pressure of fluid;

Pm - potential of mass forses;

Re - Reynolds number;

R - radiuse of curvature for curve $z=z(l)$;

ρ - radiuse of curvature for curve $y=y(s)$;

ρ_{yl} - radiuse of curvature for curve orthogonal to curve s;

ρ_f - density of fluid;

U - speed of fluid;

(x,y,z) - Cartesian coordinate system;

(l,yl,z^*) - special curvilinear coordinate system;

Tl - viscouse term in direction l;

Tyl - viscouse term in direction yl;

Tz^* - viscouse term in direction z^*;

REFERENSES

1. Vysotsky L.I. Geometrical Form of Navier-Stoks Equations and it's Application to Reverse Tasks. VINITI (1991)(in Russian).

WHAT'S THE MATTER WITH TRANS-CRITICAL FLOW?

A B SWANN

School of Civil Engineering, The University of Birmingham, UK.

ABSTRACT

The problem of computing surface profiles which pass through the critical depth is outlined. An equation is then presented based on a control volume analysis of the flow in a channel which is able to model sub-, super- and trans-critical open channel flows. The equation is used to generate a surface profile including a hydraulic jump for a typical gradually varied flow problem.

INTRODUCTION

Techniques used to predict water surface elevations in steady open channel flows generally make use of the St. Venant equations with the time derivatives set to zero. The resulting mathematical model takes the form of an ordinary differential equation which describes the rate of change of depth h with position x at any given channel section. The equation may be written as

$$\frac{dh}{dx} = \frac{S_0 - S_f}{1 - F_n^2} \qquad (1)$$

where S_0 is the bed slope, S_f the friction slope and F_n the Froude number.

Various methods have been devised for solving this equation (eg. Subramanya, 1982). Of these the most popular are probably the numerical methods since these are particularly suitable for computer application. The abiding problem is that any computational method based on equation (1) will break down when faced with a free surface which tries to cross the critical depth. That this is necessarily so is easily shown by considering the finite difference form of the equation. If a finite increment Δx is pre-defined then the resulting change in depth Δh will be given by

$$\Delta h = \left\{ \frac{S_0 - S_f}{1 - F_n^2} \right\} \Delta x \tag{2}$$

When the free surface is at the critical depth the condition $F_n = 1$ will apply which when substituted in equation (2) yields $\Delta h = \infty$. Conversely if a finite increment Δh is pre-defined, then the resulting change in position Δx will be given by

$$\Delta x = \left\{ \frac{1 - F_n^2}{S_0 - S_f} \right\} \Delta h \tag{3}$$

and putting $F_n = 1$ yields $\Delta x = 0$. One way of ameliorating this problem which has been suggested (Chawdhary 1991) is to use an implicit form of the dynamic equation. This has the effect of altering the detailed numerical calculations without affecting the mathematical integrity of the results.

Since the depth derivative effectively appears twice in equation (1) an implicit version may be obtained by taking one of the derivative terms to the other side. Thus

$$\frac{dh}{dx} = S_0 - S_f + F_n^2 \frac{dh}{dx} \tag{4}$$

The equation is solved iteratively by substituting successive mean values for $\frac{dh}{dx}$ and is progressed along the channel using any convenient numerical method. If the flow is trans-critical then an energy loss, additional to the loss due to fluid friction, will be incurred (Abbott, 1979). To account for this a second iterative procedure may be required.

Mathematical models designed specifically to analyse trans-critical flow in an open channel are available based on the concept of the control volume. In this type of model, since the crossing of the critical depth occurs within the control volume, it does not directly affect the calculations. In developing models of this type it is often assumed that the section containing the trans-critical flow can be assumed to be short thus allowing friction and gravity forces to be neglected. Clearly these assumptions are not consistent with those which apply when the St Venant equations are used. Consequently, the trans-critical portion of any flow has to be treated as a distinct entity separate from the rest of the flow in the channel. When modelling a flow containing a transition from super- to sub-critical flow the control volume model has to be made to operate between the ordinary differential equation models representing the flows on either side.

The present work attempts to combine these two approaches with the aim of representing all three classes of open channel flow through the use of a single dynamic equation, thus providing a unified approach to the problem.

THE REACH-DISCHARGE MODEL

The proposed model is based on a control volume analysis of the flow in a channel reach. Consistency with ordinary differential equation models is sought by making the length of the control volume Δx comparable with the finite difference used when obtaining numerical solutions of the ordinary differential equation. Since trans-critical flows are known to give rise to unpredictable changes in flow energy, it is necessary to consider momentum rather than energy conservation when formulating the governing equation. Additionally, since the equation is intended to be a generally applicable channel flow model both fluid friction and gravity forces have to be taken into account. The result, taking the discharge Q (m^3/s) as the dependent variable, is an equation of the form

$$Q|Q| = \frac{F_g + F_H}{C_{mom} - C_{fric}} \tag{5}$$

where F_g and F_H are the gravity and net hydrostatic pressure forces respectively, and C_{mom} and C_{fric} are momentum and friction coefficients (see Swann 1995).

Figs. 1 to 3 show the relationship between mean surface slope and discharge for a typical rectangular channel having zero, sub- and super-critical bed slopes respectively. The associations between the discharge curves and the surface profiles present in the channel are indicated on the figures. The notation is based on the usual classification system for surface profiles and all profiles are represented with the exception of those arising from flows in a channel having a critical bed slope. It is found that when the bed slope is critical the discharge curves collapse onto the zero mean surface slope axis. This would appear to suggest that the critical slope surface profiles are in fact horizontal straight lines.

Superimposed on the discharge curves are curves showing the upstream and downstream Froude numbers. By arranging the $F_n = 1$ axis coincident with the zero discharge axis the graph shows the points at which either the upstream or the downstream flow changes from one regime to the other. When the upstream flow is super-critical and the downstream flow is sub-critical a hydraulic jump must exist within the reach. Conversely, if the upstream depth is sub-critical and the downstream depth super-critical then the reach must be acting as a channel control. It can be observed that the discharge curve does not suffer any disturbance as a result of the presence of the jump which suggests that the model is stable in the presence of a trans-critical flow and is capable of representing trans-critical as well as sub- and super-critical flows. The curves also provide additional insight into the properties of the standard surface profiles.

EXAMPLE 1

The reach-discharge equation can be used to obtain surface profiles either by computer or by semi-hand calculation. Fig 4 shows surface profiles for a flow of $11.21(m^3/s)$ in a horizontal rectangular channel $5(m)$ wide and $64(m)$ long terminating in a free overfall. Manning's n for the channel is 0.02.

Since the numerical procedure used to generate this surface profile progresses from the upstream end of the channel the existence of the free overfall can only influence the calculation retrospectively. It is therefore necessary to trace all possible surface profiles and to select the one which gives critical depth at the downstream end of the channel.

The advantages claimed for the method are as follows.
- The same equation is used throughout.
- The solution progresses in a consistent direction which can be either upstream or downstream
- Friction and gravity forces are accounted for when the flow is trans-critical.
- By analysing the relative magnitudes of the four terms in equation (5) it is possible to determine on which type of surface profile the solution algorithm is working (Swann 1995).

The significance of this last point is that it allows the algorithm to adapt automatically to the particular reach-discharge curve which it is attempting to negotiate.

The disadvantages of the method are
- If the flow includes an hydraulic jump then the method can only determine in which reach the jump will occur rather than its precise position.
- It is necessary to trace more than one surface profile.

CONCLUSIONS

A single equation capable of modelling sub-, super- and trans-critical flows has been introduced. Characteristic discharge curves for a typical open channel have been presented and these have been related to the surface profiles known to occur in open channels. It has been found that all the profiles in the standard classification system are represented on the discharge curves. Sections of the discharge curves representing the hydraulic jump have also been identified. Finally the equation has been used to model a typical steady open channel flow problem including a hydraulic jump.

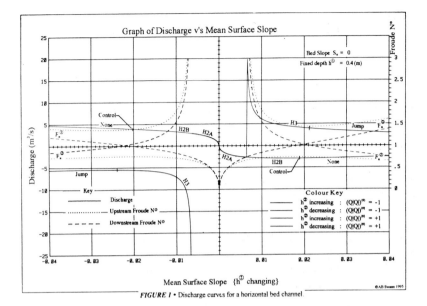

FIGURE 1 • Discharge curves for a horizontal bed channel.

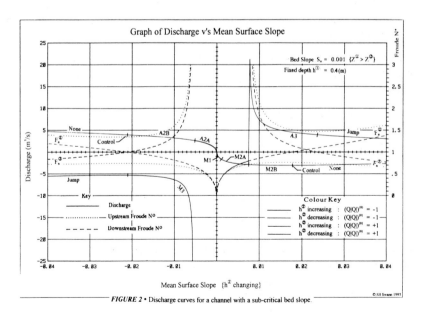

FIGURE 2 • Discharge curves for a channel with a sub-critical bed slope.

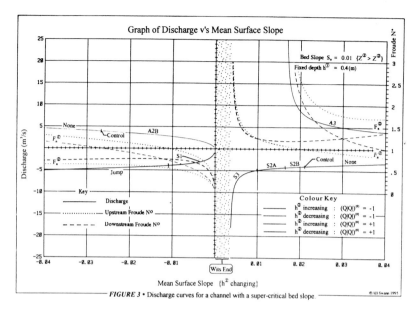

FIGURE 3 • Discharge curves for a channel with a super-critical bed slope.

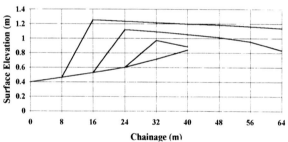

FIGURE 4 • Surface profiles for Example 1

REFERENCES

1. **Abbott, M.B.**, 1979, "Computational hydraulics", Pitman, London
2. **Chawdhry, K.S.**, 1991, "On the solution of implicit first-order differential equations", *MSc thesis*, The University of Reading, UK
3. **Subramanya, K.**, 1982, "Flow in Open Channels" Vol. 1, pp 132-166, Tata-McGraw-Hill, India.
4. **Swann, A.B.**, 1995, "A unified equation of steady non-uniform open channel flow", *MPhil thesis*, The University of Birmingham, UK.